The Bromeliads

The Bromeliads

Text by Lyman B. Smith
Paintings by Margaret Mee

South Brunswick and New York: A. S. Barnes and Company
London: Thomas Yoseloff Ltd

Library of Congress Catalogue Card Number: 68-27221

A. S. Barnes and Co., Inc.
Cranbury, New Jersey 08512

Thomas Yoseloff Ltd
108 New Bond Street
London W. 1, England

SBN—498-06887-0
Printed in the United States of America

Introduction

Originally the plates in this album were prepared by Mrs. Margaret Mee as illustrations for the *Flora Brasilica,* an encyclopedic work on all of the flowering plants of Brazil, founded by Dr. Frederico Carlos Hoehne, director of the Instituto de Botânica in São Paulo. Between 1940 and 1955, a number of well-illustrated volumes were published, but by the sixties, when Margaret Mee's paintings were ready for use, printing costs had risen far above the budget allowance for colored plates.

In 1965, I tried to break the impasse by buying twenty of her paintings for the Smithsonian Institution and afterward the present director of the Instituto, Dr. Alcides R. Teixeira, lent me thirty more for as long as needed. My plans for publication by the Smithsonian Institution were not successful, but I did find a publisher who agreed that there were enough people who appreciated beauty in plants beautifully painted to make this a worthwhile publication.

Of the paintings shown here, numbers 1 to 11 and 13 to 16 are by courtesy of the Smithsonian Institution, while number 12 and numbers 17 to 29 and 31 and 32 are by courtesy of the Instituto de Botânica of São Paulo. Plate number 30 is from a painting given to my wife by Mrs. Mee. Photography for preparation of color plates has been done by Henry B. Riblet of Kensington, Maryland.

So much for general history and statistics. Now as you look at these pictures, remember that they achieved their artistic effect and scientific accuracy because Margaret Mee was not satisfied to drag a dying plant back to the comfort of her studio nor to reconstruct from a dead one. Instead, she met them where they lived and painted them there, evolving a highly successful technique through trial and error and undergoing some hair-raising experiences in the process.

Contents

Introduction 5

About Bromeliads 11

Bibliography 14

1. Encholirium spectabile 15
2. Pitcairnia decidua 16
3. Tillandsia dura 18
4. Tillandsia stricta 20
5. Tillandsia streptocarpa 21
6. Tillandsia linearis 22
7. Vriesea billbergioides 23
8. Vriesea sceptrum 24
9. Vriesea gradata 25
10. Vriesea erythrodactylon 26
11. Vriesea incurvata 27
12. Vriesea inflata 28
13. Vriesea carinata 29
14. Vriesea ensiformis variety bicolor 31
15. Vriesea guttata 32
16. Vriesea simplex 33
17. Neoregelia hoehneana 34
18. Nidularium burchellii 35
19. Nidularium billbergioides 37
20. Nidularium seidelii 38
21. Wittrockia superba 39
22. Hohenbergia horrida 41
23. Aechmea orlandiana 42
24. Aechmea fosteriana 44
25. Aechmea caudata 45
26. Aechmea cylindrata 46
27. Aechmea vanhoutteana 47
28. Quesnelia humilis 48
29. Quesnelia liboniana 50
30. Quesnelia marmorata 51
31. Billbergia amoena variety penduliflora 52
32. Portea leptantha 54

Plates 55

The Bromeliads

About Bromeliads

DEFINITION

First of all, what is a bromeliad? We are often able to recognize them by their leaves alone without quite realizing why. Partly it is because we have seen something similar in bloom that was an obvious bromeliad, but partly it is the recognition that if the parallel-veined leaves are covered with scales, or form a rosette holding water, there is not much else that the plant can be. For good measure the rosette-leaves also show scales if you look closely at their bases.

When they do flower, the inflorescence, or group of flowers, is tremendously varied, but the form of the flower itself is relatively uniform. The inflorescence may be terminal or lateral; it may consist of an immense loose panicle, a dense head or spike, or even a single flower. The flower itself is regular or nearly so and divided into three's as in a lily, but unlike a lily it has sepals and petals very different from each other and an ovary that is inferior in many cases.

GEOGRAPHY

The next question is where to find bromeliads. With the exception of a single species in West Africa the *Bromeliaceae* are the great American tropical family. They range from coastal Virginia in a great arc to Lower California and thence south to central Argentina and Chile. Within that great sweep they are to be found anywhere and everywhere. On both coasts there are species that grow within the reach of salt spray, and in the extreme coastal deserts of Peru and Chile they flourish without roots where even cacti are unable to grow. With scarcely any change in form (see Plate 3, *Tillandsia dura*) they climb to the top of the rainforest where they are alternately drenched and scorched. Other bromeliads have entered the forest floor, where their broad thin leaves strain to catch the weak and broken light. Both types follow the forest from the coast to the tree line, where the treetop types continue and are joined by the alpine types. In the Andes *Puya* goes from sea level in Chile to 14,650 feet (4400 meters) in Bolivia. In fact there is practically nowhere in this great area where some bromeliad cannot be found.

However, the number of species in any given area varies tremendously. Most of the family area in the United States is occupied by a single species, *Tillandsia usneoides,* the Spanish moss, and a closely similar situation obtains at the southern limit of the family. The great wealth of species is in the west from Mexico to Peru and in eastern Brazil; the Amazon Basin proves surprisingly poor. Even in the best regions there is considerable variation, with more species at the higher altitudes. Thus to find a maximum of bromeliad species one should explore the rainforest of the mountain slopes of the east or the west; but here another variable enters.

In the west you might find a larger total of bromeliads, but if you are interested in ornamental species, nothing approaches eastern Brazil.

So far there seems to be no explanation for this situation, but the fact remains that genera that are almost completely ornamental are largely or wholly limited to eastern Brazil, while others—like *Tillandsia, Vriesea,* and *Aechmea*—have more ornamentals there than elsewhere. Hence the pictures of our album, although originally prepared for scientific publication, serve equally well for horticulture or art.

CLASSIFICATION

Bromeliads are easily divisible into three subfamilies—unless you wander off into the never never land of the Guayana Highlands where everything is topsy turvy. The first subfamily is the *Pitcairnioideae.* It is characterized by capsular fruits with winged or tailed seeds. That is, the seed-appendages are all of a piece and not split into hairs as in the next subfamily. The capsule, when ripe, splits neatly into three equal valves with lustrous dark chestnut linings. The ovary is mostly superior, that is, it stands above the sepals and petals. The *Pitcairnioideae* are illustrated by our first two plates, *Encholirium spectabile* and *Pitcairnia decidua.* With very rare exceptions, which do not concern us here, the plants grow on soil or rock and have not developed tank rosettes holding water. Most members of the subfamily have leaves that are more or less spiny; but occasionally one encounters a completely spineless species like *Pitcairnia decidua.* The only other genus that one is likely to meet in eastern or central Brazil is *Dyckia.* It is easily recognized by its broad orange to yellow flowers with a basal tube of fused petals and stamens.

The second subfamily, the *Tillandsioideae,* is characterized by capsular fruits with plumed seeds and leaves that never have any spines. Here we have both types of leaf arrangement and function. In the simple and presumably primitive type the leaves do not form a tank but absorb water and food through their scales, which are found over their whole surface. In Brazil, *Tillandsia* is wholly of this type. In the more complex type, the leaf has developed a division of labor that makes the tank rosette possible. The leaf-sheaths have broadened and bulged outward while their margins remain so tightly stuck that each sheath forms a water-tight pocket. Meanwhile the blade too has broadened and has lost most or all of its scales on the upper surface. Instead of absorbing water, the blade channels nearly all of it down into the tank, and at the same time the removal of the scaly covering gives the chlorophyll more light so that the leaf adapts to shade and enables the plant to live beneath the forest canopy instead of on top of it. In Brazil, *Vriesea* is of this second type. Outside Brazil, the leaf distinction breaks down and one has to fall back on the basic character of the simple petal of *Tillandsia* and the appendaged petal of *Vriesea* (see Plate 7, *Vriesea billbergioides*). The other genera of the *Tillandsioideae* are unknown or unimportant in Brazil.

The third subfamily, the *Bromelioideae,* is characterized by a berry type of fruit with unappendaged seeds. Its ovary is inferior, that is, it appears to stand below the sepals and petals (see Plate 31, *Billbergia amoena* var. *penduliflora*). The leaves are usually spiny, sometimes ferociously so. Arrangement in tank rosettes is the general but not invariable rule. The form of the rosette is very variable, ranging from an open bowl type common in many genera to the tall slender vases of certain species of *Aechmea, Quesnelia,* and *Billbergia.*

Dividing the *Bromelioideae* into genera is a tough problem on which even the professionals disagree strenuously, so it is small wonder if the amateur is perplexed. However, there are some scattering rays of light in this jungle. About a quarter of the subfamily's 600 odd species are in the largest genus, *Aechmea.* If we qualify this one in four chance by the characters of a definite scape and flowers not over an inch long, we can reverse the odds to better than three in four, but after that one must either dissect the flower and run down the genus in a technical key or compare with figures (see Graf, *Exotica*). In the latter case one generally identifies genus and species simultaneously by matching plant and figure. However, there are numerous pitfalls and it is well to check the flower with a technical description.

The smaller genera of the *Bromelioideae* are each distinguished on a combination of characters based on the form and arrangement of the flowers. *Neoregelia, Nidularium* and *Wittrockia* have the flowers densely massed together and usually surrounded by large bracts or leaves. Of these, *Neoregelia* has pedicelled flowers and unappendaged petals, *Nidularium* has sessile flowers and unappendaged petals, and *Wittrockia* has subsessile flowers and appendaged petals. There are partial secondary characters by which some species can be identified as in the brightly colored inner leaves of many neoregelias, but nothing that can take the place of dissecting a flower.

In the remaining genera of Brazilian *Bromelioideae* shown here the flowers are in obviously branched inflorescences or at least are not surrounded by large bracts or leaves and all have appendaged petals. *Hohenbergia* differs from *Aechmea* and the others in its branched inflorescence of from few to many small cones. In *Aechmea* there are species with flowers in a cone or with a branched inflorescence, but never the combination. On the other side of *Aechmea* we have *Quesnelia* and *Billbergia,* distinguished by their large flowers and soft-pointed sepals, and *Portea* by its slenderly pedicellate flowers and spine-tipped sepals. There are thirteen other Brazilian genera in this subfamily, but since they are not illustrated here we will not consider them.

Much use has been made of the form of the pollen grains to distinguish genera, but identification can be made without it and recent research shows that it is neither so simple nor so infallible as the first random samplings indicated.

Classification to species within each genus is again largely by form and grouping of the flowers, but the form of the bracts becomes much more important while the form and coloration of the leaves is diagnostic in a few species and helpful in nearly all. This is easily verified by comparing our plates of *Vriesea* or *Aechmea.*

LEAF-SCALES

Of all the characters that distinguish bromeliads as a family, only the leaf-scales are unique. The other characters must be used in combination because each one by itself occurs in many other families. That is not to say that no other families have scales on their leaves—a great many do. However, in all these other families the scales are only a protection against loss of moisture, and absorption of water remains a function of the roots. In the *Bromeliaceae* the leaf-scale is not merely an outgrowth of the epidermal layer, but it has a complex organization with a column of stalk cells beneath it that penetrates the leaf and pumps in the water collected by the scale.

In this respect, bromeliads make an interesting contrast with two of their tree-top neighbors. Some epiphytic orchids have developed chlorophyll-bearing roots and lost their leaves, and most cacti have stems that function as leaves; but bromeliads have leaves that function as roots, while the roots have become mere fastening clamps and finally have disappeared altogether.

The leaf-scales are constant in form within a species but may show different types on the sheath and blade. As between species, there is considerable difference; but on the other hand there are some similarities that are far reaching. A single scale by the regularity of its design indicates a member of the *Tillandsioideae.* There are four identical central cells with rings of cells radiating outward and doubling in number as they go. In the other two subfamilies the scales are very irregular aggregations of cells with no clear-cut center or design. In both cases the scale may be nearly circular, but in the *Pitcairnioideae* and *Bromelioideae* they become stellate in many species by even prolongation in several directions, while in the *Tillandsioideae* there is often a single retrorse prolongation by the outermost ring of cells, giving the leaf a shaggy appearance.

CULTIVATION

Many if not most of the people who use this album will be interested primarily in cultivating bromeliads, but this is a special subject better handled by the references in the bibliography below. This album aims rather to provide good pictures for identification by comparison or for pure ornament in themselves.

It should be noted, however, that bromeliads are among the easiest of plants to cultivate, the main problem being one of space due to their complete disregard for population control.

Bibliography

GENERAL:

The Bromeliad Society Bulletin, by membership: Mrs. Jeanne Woodbury, 1811 Edgecliff Drive, Los Angeles, California, 90026. Something for everyone from the beginner to the professional.

Bromeliads in Color and their Culture. Editor, Bromeliad Society, 647 South Saltair Avenue, Los Angeles, California, 90049. As titled.

Bernardi, L., & Robert, P.-A. *Fleurs Tropicales.* Delachaux et Niestlé, 32 rue Grenelle, Paris VII, France. Page 300: *Une famille millionaire avec piscine et zoo privés: les Bromeliacées.* You can enjoy it with even your high school French and the pictures speak all languages.

Foster, M. B. & R. *Brazil Orchid of the Tropics.* The Ronald Press Company, 15 East 26th Street, New York, New York, 10010. Despite the title, almost wholly about bromeliads.

TECHNICAL:

Mez, C. Bromeliaceae. *In Engler, Das Pflanzenreich, IV. Familie 32.* 1934-35. The latest complete monograph of the family.

Smith, Lyman B. *The Bromeliaceae of Brazil,* Smithsonian Miscellaneous Collections, volume 126. 1955. The latest complete treatment of the Brazilian species.

HORTICULTURE:

Graf, A. B. *Exotica,* edition 3. Roehrs Company, Rutherford, New Jersey. Photographs of all the commonly cultivated bromeliads and a great many of the rarer ones, Edition 4 in preparation.

Wilson, R. G. & C. Bromeliads in Cultivation. Edwin A. Menninger, Stuart, Florida. First of two volumes with genera in alphabetical order. Many colored figures with big coverage of *Aechmea.*

1. Encholirium spectabile

Encholirium means "sword-lily" and the ramrod-straight six-foot flowering stalk adds to the military appearance of this hardy inhabitant of the arid northeast, Brazil's least hospitable area. Dr. Von Martius, the famous botanical explorer of Brazil, discovered the genus in 1819 on the Ilha do Fogo (Fire Island) near Joazeiro in northern Bahia. It grows on bare rock that is but slightly more sterile than the soil around it and makes a good living of it in the bargain, often forming dense impenetrable masses; and according to the early English explorer, George Gardiner, it sometimes produces flower stalks 15 feet tall. The more modest plant depicted by Margaret Mee was found on the Serra do Vento in Pernambuco, Brazil's easternmost state.

Appropriately enough for our first number, *Encholirium* is a wholly Brazilian genus and *E. spectabile* is its oldest and most widespread species, being found all over the northeast and as far west as the new capital of Brasilia. There it grows between the jagged spurs of limestone outcrops, fully exposed to the blazing sun. The eight species that have been added to *Encholirium* have much the same appearance and are of very retricted range.

The modestly colored yellow-green flowers and formidable spiny leaves of Dr. Martius' "spectacular sword-lily" are not such as to inspire a craze among people with limited growing space, but in a cactus garden it finds itself with old neighbors and gives a strong dash of diversity to the whole.

Encholirium Martius ex Schultes filius in Roemer & Schultes, *Systema Vegetabilium,* volume 7, part 2, p. lxviii, 1233. 1830. Coarse perennial herbs. Leaves rosulate at the ends of short stout stems or branches, coarsely spinose-serrate. Inflorescence mostly simple, rarely few-branched. Flowers pedicellate, perfect. Sepals free, symmetric, glabrous. Petals free, unappendaged, narrower than the sepals. Stamens mostly exceeding the petals, the filaments free. Ovary wholly superior, glabrous. Seeds numerous, each with a single dorso-apical wing much extended backward.

Encholirium spectabile Martius ex Schultes filius in Roemer & Schultes, *Systema Vegetabilium,* volume 7, part 2, p. 1233. 1830. PLANT flowering 2-5 meters (6-15 feet) high. LEAVES 60 cm. long, glabrous above, lepidote beneath, serrate with teeth over 10 mm. long. SCAPE elongate, stout; scape-bracts remote, deciduous with age. INFLORESCENCE densely racemose, rarely branched (M. B. Foster), 40-90 cm. long, glabrous. FLORAL BRACTS linear, entire, recurved, usually much shorter than the pedicels. FLOWERS 25 mm. long, yellow-green; pedicels slender, 8-14 mm. long. SEPALS ovate, obtuse, 6-8 mm. long. PETALS lanceolate, 20 mm. long, 5 mm. wide. STAMENS with anthers 8 mm. long. STYLE slightly shorter than the ovary.

2. Pitcairnia decidua

The name "Pitcairnia" conjures up visions of Pitcairn Island and *Mutiny on the Bounty,* but actually the genus was named in honor of William Pitcairn, a London physician, and comes no nearer to Pitcairn Island than the coast of Mexico or Peru. Quite unlike its relative, *Encholirium,* its center of development is in northwestern South America and relatively few of its 260 odd species are native to Brazil. However, these last form an interesting link with the one "unAmerican" bromeliad, *Pitcairnia feliciana,* of West Africa.

At first glance, *Pitcairnia decidua* closely resembles *P. feliciana* in size and form; but as its specific name implies, its leaf-blades are deciduous while those of *P. feliciana* are persistent, withering only in extreme age and then shredding away irregularly. As Margaret Mee has emphasized by her separation of leaves and flowers, *Pitcairnia decidua* rarely if ever displays both at the same time. During the growing season, corresponding to our northern summer, its grass-like leaves develop, and then as drier weather approaches, the base of the blade is neatly severed as if by a razor and the floral shoot begins to develop. This characteristic is rare in bromeliads and is almost wholly confined to *Pitcairnia,* where most of the deciduous-leaved species occur in Mexico in a climate of much more marked seasonal change.

The plate illustrates another character that is almost wholly limited to *Pitcairnia* and much more common there than deciduous leaves, namely the zygomorphic or irregular corolla. Notice that the tips of the petals are in a plane above the stamens instead of being evenly placed on a circle around them. This irregularity is very slight when compared to the petals of an orchid, but even though the pitcairnia's petals have equal dimensions, each has an individual curvature necessary to bring it next to the others over the stamens. A closer comparison than the orchid flower would be with that of a gladiolus.

Margaret Mee's plant was cultivated in the gardens of the Instituto de Botânica in São Paulo and is a descendant of the original plant collected on sun-drenched rocks at Itabapoana in Espirito Santo. Apparently its softer living in the garden has changed neither its habits nor its appearance appreciably.

Pitcairnia L'Héritier, *Sertum Anglicum* 7. 1788. Stemless or rarely caulescent herbs, nearly always terrestrial. Leaves fasciculate or imbricate along a stem, in 2 or more ranks, entire or spinose-serrate, the sheaths small, sometimes bulbous-thickened, the blades from linear to broadly lanceolate with a distinct petiole or narrowly triangular, sometimes two or three types in one species, rarely deciduous along a regular transverse line. Inflorescence simple or branched. Flowers pedicellate to nearly sessile. Sepals free, acute or obtuse. Petals free, usually somewhat irregular, naked or bearing a single scale near the base. Stamens elongate, from shorter than to slightly exceeding the petals, the anthers linear. Ovary from almost wholly superior to completely inferior, the style filiform, about equaling the petals, the ovules numerous, bearing entire appendages. Fruits dehiscent or indehiscent, the seeds winged or bicaudate or very rarely naked.

Pitcairnia decidua L. B. Smith in *Arquivos de Botânica do Estado de S. Paulo,* nova série, volume 1, fasciculo 5, p. 110, pl. 114. 1943. PLANT stemless, flowering to 4 dm. high, white-flocculose at first. LEAVES many in a bulbiform rosette, entire; sheaths ovate, dark castaneous; blades dimorphic, the outer filiform, very short, soft, persistent, the others deciduous along a straight transverse line, linear, filiform-attenuate, 6 mm. wide at base. SCAPE erect, 3 mm. in diameter, dark when dry; scape-bracts erect, few but much exceeding the internodes, the lower foliaceous, the others ovate with a long filiform blade. INFLORESCENCE simple, 2 dm. long, sublax. FLORAL BRACTS about equaling the pedicels, ovate or lanceolate, attenuate. FLOWERS suberect; pedicels slender to 3 cm. long. SEPALS narrowly triangular, acuminate, ecarinate, to 22 mm. long. PETALS about 4 cm. long, naked, red. OVARY 4/5 superior; ovules long-caudate.

3. Tillandsia dura

Tillandsia and *Bromelia* are the two original genera of bromeliads published by Linnaeus in 1753 and both are named in honor of botanists. However, the Father of Systematic Botany had an uncomfortably sharp sense of humor and he named *Tillandsia* for a student who was so seasick on his first voyage that he went 200 miles out of his way to return home by land. The student thus became *Tillands* or "the landlubber." In the end though, the joke was on Linnaeus, because he thought that the bromeliad was also averse to water in having scales to shed it; actually they do the reverse.

The two tillandsias published in 1753 have now grown to some 350 species, making it the largest and most varied genus in the *Bromeliaceae*. The giants of the genus and the great majority of the species are western, leaving a handful of medium to tiny ones in Brazil. These are sun-loving plants with narrow tapering leaves, and they contrast strongly with their nearest relatives in *Vriesea*, which are shade-loving and have broad, thin leaves.

Tillandsia dura is a member of the subgenus *Allardtia*, which has stamens only a little shorter than the narrow petals. This appears to be the primitive part of *Tillandsia*, centering in the northern Andes with the other subgenera radiating out from it; significantly the nearest relatives of *Tillandsia dura* are in Bolivia and Peru.

Evidently *Tillandsia dura* is a tree-top species, for it is collected chiefly after it or its tall host have fallen to the forest floor. Margaret Mee obtained the specimen for her painting from the forests of the Serra do Mar (Coast Range) of São Paulo, but the species is found also in Rio de Janeiro and as far south as Santa Catarina. Mez had an excellent black and white plate (No. 108) in the *Flora Brasiliensis*, but the present portrait is the first one in color.

Tillandsia Linnaeus, *Species Plantarum*, p. 286. 1753; *Genera Plantarum* edition 5, p. 138. 1754. Stemless to long-caulescent herbs of highly variable habit. Leaves rosulate or fasciculate or distributed along a stem, in two or more ranks, entire, triangular or sometimes ligulate. Scape usually distinct. Inflorescence variable, usually compound with distichous-flowered spikes, sometimes simple and polystichous-flowered from the reduction of the spikes to a single flower each, and rarely the whole inflorescence reduced to a single flower. Floral bracts usually conspicuous. Flowers mostly short-pedicellate, actinomorphic or radially symmetric. Sepals mostly symmetric, free or equally or posteriorly connate. Petals wholly free, twisted to the right, unappendaged. Stamens of various lengths in relation to the petals and pistil, free or those of the second series shortly adnate to the petals. Ovary superior, glabrous; ovules usually many and caudate. Capsule septicidal; seeds erect, slenderly cylindric or fusiform, bearing a white straight plumose appendage at base.

Tillandsia dura Baker, *Handbook of the Bromeliaceae,* p. 168. 1889. PLANT 2-4 dm. long; stem almost always simple, from very short to 14 cm. long. LEAVES very numerous, polystichous, strict or somewhat arching and often secund, 15-23 cm. long, densely and finely appressed-lepidote, subglabrous above with age; sheaths ovate, scarcely more than 1 cm. long, passing into the blade; blades narrowly triangular, to 11 mm. wide but usually much narrower, subulate-attenuate, rigid. SCAPE distinct but usually much covered by the leaves, terminal, slender, erect or ascending; scape-bracts densely imbricate, ovate, thin, densely and finely appressed-lepidote. INFLORESCENCE always simple and distichous-flowered, linear, acute, strongly complanate, 7-13 cm. long, 10-14 mm. wide, 14-26 flowered; rhachis slightly geniculate, angled, glabrous. FLORAL BRACTS erect, strongly imbricate but so narrow as partially to expose the rhachis, ovate, acute, apiculate, 17-20 mm. long, exceeding the sepals, subchartaceous, strongly nerved, densely appressed-lepidote toward apex, rounded or carinate above, straight or slightly incurved. FLOWERS subsessile. SEPALS free or equally short-connate or at times the posterior ones connate for about half their length, 8-11 mm. long, oblong-lanceolate, acute, strongly nerved, glabrous. PETALS narrow, blue, 17 mm. long. STAMENS barely included; anthers linear, acute, 5 mm. long. PISTIL exceeding the stamens; ovary slenderly ovoid.

4. Tillandsia stricta

The great majority of tillandsias with a simple cone-like inflorescence like that of *Tillandsia stricta* belong to the subgenus *Anoplophytum,* the only subgenus centering in Brazil. Technically the only difference from the subgenus *Allardtia* is its shorter stamens that have shrunk back into the throat of the corolla.

Although some of its near relatives are found only in Rio de Janeiro and the immediate vicinity, *Tillandsia stricta* is a true "bandeirante" or Brazilian pioneer whose vitality has extended it north to Trinidad and south to Argentina. It colonizes any convenient tree and is a common sight on telephone wires. Adda Abendroth has it in her garden at Teresópolis in the heart of the famed Organ Mountains above Rio and writes that it grows on trees in the forest and invades gardens and trees in public squares. She has been able to distinguish several forms or varieties, but Margaret Mee's plant from São Paulo is very close to the original form depicted in plate 1529 of the *Botanical Magazine* back in 1813.

Its foliage is not spectacular, but *Tillandsia stricta* compensates by flowering frequently and rather irregularly. Of the several forms distinguished by Dona Adda only one shows any regularity. She has also noted that like many other bromeliads, it is visited and presumably pollinated by hummingbirds.

Tillandsia stricta Solander in *Botanical Magazine,* volume 37, plate 1529. 1813. PLANT flowering 10-22 cm. high; stem very short. LEAVES many, rosulate, sometimes secund, 6-18 cm. long, from much shorter than the inflorescence to exceeding it, covered with appressed cinereous scales; sheaths narrow and obscure; blades very narrowly triangular, long-attenuate, 4-11 mm. wide at base. SCAPE erect to decurved, slender, usually well developed but sometimes short; scape-bracts imbricate, the lower subfoliaceous, the upper elliptic with linear blades, membranaceous, lepidote. INFLORESCENCE simple, dense, or lax toward base, 2-7 cm. long, 15-35 mm. in diameter. FLORAL BRACTS erect to spreading, elliptic, membranaceous, yellowish white to rose, the lower exceeding the flowers, lepidote toward apex, filiform-laminate. SEPALS lance-ovate, acute, 9-13 mm. long, equally connate for 2-4 mm., membranaceous, glabrous. PETALS 15-24 mm. long, blue or purple; blades narrow, obtuse or emarginate. STAMENS about equaling the petal-claws; filaments plicate. CAPSULE slenderly cylindric, to 4 cm. long.

5. Tillandsia streptocarpa

Tillandsia streptocarpa is a member of the subgenus *Phytarrhiza,* which is distinguished by its short stamens and even shorter pistil and by its broad showy spreading petal-blades. This subgenus evidently started in the central Andes of Ecuador and Peru and by adapting to drier and drier conditions pushed southeast into Paraguay and Argentina, then swung north into Brazil's arid "Nordeste."

It was here in the state of Pernambuco that Margaret Mee made several trips with the help of Dr. Dardano Lima and found the plant shown in this plate. Notice how twisted the leaves are, enabling them to grip branches and help the roots to fasten the plant. Some related species go a step further in the direction of Spanish moss and lose their roots after they are well established.

Like many xerophytic bromeliads, *Tillandsia streptocarpa* grows just as well on rocks as on trees. Far to the south of Pernambuco in the state of Paraná there is a curious sandstone formation that looks so remarkably like a medieval walled city that it is called "Vila Velha." As one comes close to its vertical walls one finds them covered with thousands of plants of *Tillandsia streptocarpa* in all stages of development—and nearly all out of reach.

Tillandsia streptocarpa Baker in *Journal of Botany,* volume 25, p. 241. 1887. PLANT flowering 1-5 dm. high; stem 1 dm. long to almost lacking. LEAVES 8-50 cm. long, covered with slender spreading cinereous scales; sheaths densely imbricate along the stem or subrosulate, broadly ovate or suborbicular, 1-2 cm. long; blades spreading or recurved, linear-triangular, long-attenuate, to 15 mm. wide at base, more or less involute-subulate, usually recurved at apex and grasping the branch on which it rests. SCAPE erect, slender, nearly glabrous; scape-bracts many, imbricate, lanceolate, acute, densely lepidote, nerved, the lowest with slender foliaceous blades. INFLORESCENCE bipinnate from 2-12 spikes or rarely simple; primary bracts erect, like the upper scape-bracts, much shorter than the spikes; spikes linear to elliptic, arched-spreading from an erect densely bracteate sterile base, complanate, 3-12 flowered; rhachis flexuous, slender, angled, soon glabrous. FLORAL BRACTS lanceolate, acute or obtuse, slightly shorter than the sepals, subpapyraceous, nerved, lepidote or glabrous. FLOWERS erect. SEPALS free or very short-connate, oblong, broadly acute or obtuse, 10-13 mm. long, subcoriaceous, glabrous. PETALS *ca.* 25 mm. long, the blades spreading, very broadly obovate, blue or purple. STAMENS deeply included, exceeding the pistil. CAPSULE cylindric, slender, 3-5 cm. long.

6. Tillandsia linearis

Tillandsia linearis is the ugly duckling of the bromeliads with a bundle of stringy leaves half hiding a single small spike. Then one fine day it blossoms with three tremendous round lilac-blue petals to each flower. At this point it is easy to see its close relationship to *Tillandsia streptocarpa,* for although the habits are radically different, the flowers are nearly identical except in size.

In comparison with its wide-ranging relative, *Tillandsia linearis* is a homebody, being limited to Goiás and southeastern Brazil from the vicinity of Rio to Paraná. In the southern part of its range it is found on the lordly pinheiros or araucarias that dominate the landscape around Curitiba, while in the north it grows in a typical rainforest of broad-leaved trees. Margaret Mee's plant comes from Parati in western Rio de Janeiro, and may well be from the very spot where old Frei Vellozo discovered it nearly two hundred years ago.

Frei Vellozo was a Franciscan monk who was commissioned to write a *Flora Fluminensis,* that is a flora of the vicinity of Rio de Janeiro. Beginning in 1782, he worked steadily for eight years collecting and preparing for his flora. Unfortunately he did not live to see its publication as a small volume of text in 1825 and eleven great volumes of plates in 1835.

Vellozo gave no locality for his *Tillandsia linearis* but did cite Parati for a number of other species, so that it is a distinct possibility. His description leaves one in doubt but his rather crude plate shows all the essential features even to details of the flower. For many years Vellozo's work was largely ignored by the foreign botanists studying the flora of Brazil and later collections of his *Tillandsia linearis* were redescribed and then confused with *T. setacea* of the West Indies, but in 1901, Mez discovered the error and corrected it.

Tillandsia linearis Vellozo, *Flora Fluminensis,* p. 133. 1825, and *Icones,* volume 3, plate 128. 1835. PLANT stemless, flowering 13-25 cm. high. LEAVES about 10-20, erect or suberect, subfasciculate, to 38 cm. long, always exceeding the inflorescence, completely covered with small appressed cinereous scales; sheaths evident, triangular; blades linear, very slender, filiform-attenuate, 1-2 mm. thick at base. SCAPE erect, very slender; scape-bracts imbricate and wholly covering the scape, the lower ones subfoliaceous, the upper lanceolate, acute. INFLORESCENCE simple, narrowly lanceolate, strongly complanate, densely few-flowered, 30-45 mm. long, 6-10 mm. wide. FLORAL BRACTS erect, imbricate and more or less covering the rhachis, elliptic, acute, about 2 cm. long, exceeding the sepals, convex but scarcely carinate, membranaceous, strongly nerved, red, lepidote. FLOWERS sessile. SEPALS free, lanceolate, acuminate, to 19 mm. long, carinate, glabrous. PETALS with linear claw; blade suborbicular, 10-14 mm. in diameter, spreading, violet or blue. STAMENS deeply included, exceeding the pistil. CAPSULE slender, acute, to 32 mm. long.

7. Vriesea billbergioides

In *Vriesea* we find the largest and at the same time the most ornamental genus of bromeliads in Brazil, and since the great majority of ornamental bromeliads are Brazilian we can safely say that it is the largest such in the family in spite of the much greater number of species in *Tillandsia*. The basic difference between the two genera is that *Tillandsia* has no appendages on its petals while *Vriesea* has two scales at the base of each petal. In western South America *Tillandsia* and *Vriesea* imitate each other to such an extent that one can never be sure which is which without seeing fresh petals, but in Brazil the broad thin leaf of the shade-loving *Vriesea* is practically unmistakable.

Vriesea billbergioides is unique in the genus for its inflorescence that combines a pendent position with diffuse branching. Its graceful form and bright coloring make it outstanding even in a genus so rich in beautiful species as *Vriesea*. The subject of Margaret Mee's painting was collected in the Organ Mountains, the origin of a great many cultivated species of bromeliads and of other families as well. It is found also as far north as the Serra de Caparao between Minas Gerais and Espirito Santo and south to the Serra da Bocaina in São Paulo but always in the higher reaches of the coastal rainforest.

Vriesea Lindley in *Botanical Register,* volume 29, plate 10. 1843, nomen conservandum. Stemless or rarely somewhat caulescent herbs. Leaves rosulate in more than two ranks, entire, ligulate or in a few species narrowly triangular. Scape distinct. Inflorescence simple or compound with flowers two-ranked or many-ranked in both cases. Floral bracts usually conspicuous. Flowers from subsessile to long-pedicellate. Sepals symmetric, free or nearly so. Petals free, bearing two scales inside at base. Stamens included to exserted, the second series often short-adnate to the petals. Ovary superior or nearly so, glabrous; ovules usually many and caudate. Capsule septicidal; seeds erect, slenderly cylindric or fusiform, bearing a white straight plumose appendage at base.

Vriesea billbergioides E. Morren ex Mez in Martius, *Flora Brasiliensis,* volume 3, part 3, p. 534, plate 100. 1894. PLANT flowering over 5 dm. long with the pendulous inflorescence extended. LEAVES 2-4 dm. long; sheaths elliptic, large, green or purple; blades ligulate, broadly rounded and apiculate, to 42 mm. wide, pale green, glabrous, minutely lepidote beneath. SCAPE decurved or pendulous, slender; scape-bracts straight, broadly elliptic, rounded or broadly acute and apiculate, red. INFLORESCENCE laxly paniculate, 16-32 cm. long, 7-10 cm. wide, many-flowered; axes minutely lepidote; primary bracts ovate, acute, 2-3 times shorter than the racemes; racemes suberect, very laxly 4-flowered, the sterile base short, naked or rarely with a single bract; rhachis very slender, strongly geniculate. FLORAL BRACTS distichous, broadly ovate, acute or obtuse, not enfolding the base of the flower, to 15 mm. long, much exceeded by the sepals, bright red, membranaceous. FLOWERS spreading, not at all secund, yellow; pedicels about 6 mm. long. SEPALS elliptic, obtuse, ca two cm. long, carinate. PETALS ligulate, obtuse, 3-4 cm. long, bearing two oblong acute scales. STAMENS and PISTIL exserted.

8. Vriesea sceptrum

Unlike our previous example, *Vriesea sceptrum* requires nine different characters to distinguish it from its numerous close relatives. It is denser and more robust than most of them and it would make at best a very awkward and unwieldy "scepter," but obviously we have botanic license as well as poetic license in the choice of names. Also in a closely knit group of species such as this, it becomes increasingly difficult to find a name that is at all appropriate and that has not been used already.

The larger species like *Vriesea sceptrum* harbor an amazing variety of animals in their tanks, ranging from protozoa to frogs and snakes. One of the most interesting of these is the aquatic crustacean, *Elpidium bromeliarum,* which looks like a tiny shrimp swimming around with a clam shell on its back. With no discernible means of locomotion out of water it still manages to reach most bromeliad tanks. (See R. Reitz in The *Bromeliad Society Bulletin,* volume 6, p. 61, 1956.)

Margaret Mee's luxuriant specimen of *Vriesea sceptrum* came from the Campos do Jordão in São Paulo, but the species was first discovered in Caldas in southern Minas Gerais and it is one of the features of Itatiaia, the famous mountain peak between Minas and Rio that is a treasure chest of bromeliad species. Thus *Vriesea sceptrum* is one of the relatively few Brazilian vrieseas that have moved up over the divide onto the great central planalto or plateau.

Vriesea sceptrum Mez in DeCandolle, *Monographiae Phanerogamarum,* volume 9, p. 606. 1896. PLANT flowering to one meter high. LEAVES many in a funnelform rosette, 4-5 dm. long, straight; sheaths large, very broadly elliptic, covered with minute appressed scales, dark castaneous beneath; blades ligulate, acute or rounded, apiculate, 5-8 cm. wide, glabrous above, obscurely punctulate-lepidote beneath. SCAPE erect, short, stout, completely covered by the foliaceous densely imbricate bracts. INFLORESCENCE densely bipinnate, many-flowered; primary bracts erect, imbricate, more or less red, the lower subfoliaceous and more or less equalling the branches, the upper suborbicular, apiculate, and much shorter than the branches; branches suberect, 6-12 flowered, lax at least toward base, 15-25 cm. long, the lower with sterile bracts at base; rhachis geniculate, angled, glabrous. FLORAL BRACTS distichous, never secund, very broadly elliptic, strongly carinate and incurved, 25-30 mm. long, exceeded by the sepals at anthesis, coriaceous, smooth, very obscurely punctulate-lepidote. FLOWERS divergent, never secund; pedicels stout, 3-5 mm. long. SEPALS elliptic, acute, 2-3 cm. long, ecarinate, subcoriaceous, nerved, glabrous. PETALS ligulate, obtuse, 10 mm. longer than the sepals, orange, bearing two acute scales at base. STAMENS barely included.

9. Vriesea gradata

With its gracefully arching leaves and single brightly colored spike, *Vriesea gradata* is a good representative of an extremely popular and closely knit group of species. It is a relative newcomer, having been described in 1888 as a *Tillandsia,* while *Vriesea psittacina,* the oldster of the group and also the type of the whole genus, goes clear back to 1828. *Vriesea gradata* is comparatively rare also, having been found in only three more localities in nearly eighty years.

Margaret Mee discovered the most recent locality for it at Caraguatatuba, one of those sonorous Indian place names in which São Paulo abounds, as for example Paranapiacaba, Pindamonhangaba and Guaratingueta. However prosaic their meanings might be, they somehow sound as if they they belonged with the bromeliads and it is better not to try to translate them.

Why Baker ever chose the name "gradata" for this species and what he intended by it is anyone's guess. He could have referred to the floral bracts that diminish in size toward the apex of the spike, but that is true of all vrieseas and he might just as well have called it "viridis" because of its green leaves. Be that as it may, its best character, if one can so call any one of eight necessary distinctions, is its long sepals jutting out beyond the floral bracts.

Vriesea gradata (Baker) Mez in Martius, *Flora Brasiliensis,* volume 3, part 3, p. 523. 1894. *Tillandsia gradata* Baker in *Journal of Botany,* volume 26, p. 105. 1888. PLANT flowering 5-6 dm. high. LEAVES 10-15 in a dense broadly funnelform rosette, to 5 dm. long; sheaths elliptic, distinct, ca. 1 dm. long, bearing minute appressed pale brown scales, brown-tinged toward base; blades ligulate, not contracted toward base, 3-4 cm. wide, subrounded and apiculate, glabrous, wholly green. SCAPE straight; scape-bracts erect, imbricate, broadly elliptic, minutely apiculate. INFLORESCENCE simple, dense to lax at anthesis, to 27 cm. long, narrowly lanceolate, strongly compressed. FLORAL BRACTS suberect and very densely imbricate, then at least the lowest spreading and folding around the sepals and exposing the rhachis, suborbicular, apiculate and briefly incurved at apex, slightly if at all carinate, rather thin, glabrous or sparsely lepidote toward apex, red near apex and the remainder green or wholly red, 38-45 mm. long; pedicels obconic. SEPALS narrowly elliptic, obtuse, 4 cm. long, exceeding the floral bracts, thin, red or yellow. PETALS linear, obtuse, 5-6 cm. long, yellow, bearing two obtuse scales at base. STAMENS exserted.

10. Vriesea erythrodactylon

The "Red-fingered Vriesea," as Eduard Morren's appropriate title translates, is one of the unforgettables. Its single spike is a small green monster with talons dipped in blood or a file of crabs going into battle, and its leaves with their dark purple and brown sheaths are striking by themselves.

Why it is not more common in cultivation is hard to explain, as Morren had it in his gardens in Belgium as early as 1882, and it ranges from Espirito Santo to Santa Catarina, often occurring in great quantity. Although it is found at the crest of the coast range, its most exuberant growth is close to the sea. Margaret Mee found her fine plant at Caraguatatuba, up the coast from Santos, and in places along the Paraná and Santa Catarina coast it literally covers the ground.

This brings to mind a curious situation, namely the increase of individuals with the decrease in number of species. Around Rio, where the species of bromeliads reach their peak, one finds them in rather small colonies as a rule; but in Santa Catarina, where the species are much fewer, they cover the trees and the ground under the trees. The ultimate, which really does not belong in this account, is the case of *Deuterocohnia haumanii,* which is the sole cover for whole mountainsides in northwestern Argentina.

Vriesea erythrodactylon (E. Morren) E. Morren ex Mez in DeCandolle, *Monographiae Phanerogamarum,* volume 9, p. 569. 1896. *Vriesea psittacina* var. *erythrodactylon* E. Morren in *Belgique Horticole,* volume 32, p. 287. 1882. PLANT flowering 3-4 dm. high, propagating by horizontal rhizomes. LEAVES 10-20 in a broadly funnelform rosette, 2-4 dm. long; sheaths broadly elliptic, much broader than the blades, minutely brown-lepidote, brown toward base, dark purple toward apex; blades ligulate, broadly acute or rounded and apiculate, 2-3 cm. wide, green except for the dark purple apex, glabrous above, very obscurely lepidote beneath. SCAPE erect, 5 mm. in diameter; scape-bracts densely imbricate, thin, nearly glabrous, pale green, the lower polystichous, broadly elliptic, apiculate, closely enfolding the scape, the upper distichous, gradually more spreading and more like the floral bracts. INFLORESCENCE simple, densely many-flowered, oblong-obovate, strongly compressed, 10-23 cm. long, 7-10 cm. wide, glabrous. FLORAL BRACTS distichous, densely imbricate with their bases and always covering the rhachis and sepals, 6-7 cm. long, broadly ovate with a long triangular spreading apex, complicate and sharply keeled with the sides nearly flat, papyraceous, pale green, the upper ones bright red toward apex, the keel lightly sigmoid in outline with spreading base and incurved apex. FLOWERS erect with the petals all emerging on one side of the inflorescence but not truly secund; pedicels obconic, 3 mm. long. SEPALS elliptic, obtuse, ecarinate, 30-32 mm. long. PETALS ligulate, obtuse, 42 mm. long, yellow, bearing two obovate obtuse scales at base. STAMENS and PISTIL exserted. CAPSULE cylindric, acute, 3 cm. long.

11. Vriesea incurvata

Vriesea incurvata is our first "catarinense" or "barriga verde" (green belly), as the inhabitants of Santa Catarina are nicknamed. Charles Gaudichaud-Beaupre, French botanical explorer, on the second of his three globe-girdling scientific expeditions, discovered this species on the Ilha de Santa Catarina, the first part of the state to bear the name. In 1832 he need not have gone far from the docks of the present capital city of Florianópolis to find his prize and even now it is a common sight in the coastal rainforest from here to the state of São Paulo where Margaret Mee collected it.

The name "incurvata" doubtless refers to the floral bracts and does not seem particularly inspired, but remember that at the time of publication there were very few vrieseas known and that it was an extreme in this character and still remains so. It bears considerable resemblance to *Vriesea gradata* (plate 9) but is easily distinguished by its large sharply keeled yellow-margined floral bracts that completely hide the sepals. Another feature well shown by our plate is the way the flowers all issue from one side of the spike so that they are hidden from the other side.

Gaudichaud's plate makes interesting comparison, as it is a line drawing evidently prepared from dried herbarium fragments. Although it misses the beauty of our plate, it supplies all the technical details—and this is important because it has no description with it and is thus the sole basis for the new species. Another oddity, the species was published in the account of the trip on *La Bonite,* but was collected on the earlier trip on the ship *l'Herminie.*

Vriesea incurvata Gaudichaud, Atlas, *Voyage autour du monde executé 1836-37 sur la corvette La Bonite,* plate 68. 1843. PLANT flowering 3-dm. high. LEAVES 15-20 in a spreading rosette, 25-56 cm. long; sheaths broadly elliptic, much wider than the blades, pale brown toward base, green or sometimes purple toward apex, covered with minute appressed brown scales; blades ligulate, broadly acute or subrounded and apiculate, 20-45 mm. wide, sometimes slightly narrowed toward base, wholly green, sometimes with small irregular transverse nerves, glabrous above, very obscurely lepidote beneath. SCAPE erect, 5-6 mm. in diameter, much shorter than the leaves; scape-bracts strongly imbricate and enfolding the scape, broadly elliptic, apiculate, the upper red and gradually larger and more like the floral bracts. INFLORESCENCE simple, densely many-flowered, oblong or linear, acute, strongly compressed, 15-40 cm. long, 4-5 cm. wide, glabrous or nearly so. FLORAL BRACTS strongly imbricate, always covering the rhachis and sepals, erect, 45-60 mm. long, subrhombic, folded on the keel with the sides nearly flat, incurved toward apex, papyraceous and red with broad yellow membranaceous margins. FLOWERS erect with the petals all issuing from one side of the inflorescence but not truly secund; pedicels obconic, stout, 3-5 mm. long. SEPALS elliptic, obtuse, 15-27 mm. long, ecarinate. PETALS ligulate, obtuse, nearly 5 cm. long, bearing two obtuse scales at base. STAMENS and PISTIL exserted. CAPSULE cylindric, attenuate, 35-40 mm. long.

12. Vriesea inflata

The German and Austrian nobility of the nineteenth century were great patrons of botany with their greatest achievement the production of the monumental *Flora Brasiliensis*. Some even organized their own expeditions, like Emperor Maximilian and the Princes of Sachs-Coburg, and on both of these, Wawra, or rather Heinrich Wawra Ritter von Fernsee, visited Brazil as the professional botanist. There he collected in Rio and on the Organ Mountains and was the first botanist on Itatiaia. He collected *Vriesea inflata* on both occasions but passed it by as a mere variety at first and did not include it in the colored plates of either of his sumptuous reports.

Mez, in the *Flora Brasiliensis,* considered it a variety of *Vriesea incurvata,* to which it is no doubt closely related; but obviously he was working from pressed herbarium specimens or he would have been struck immediately by the difference in thickness of the spikes. Even in full bloom, the spike of *Vriesea incurvata* is as flat as a cardboard cutout while that of *Vriesea inflata,* as you might guess from the name, is like a fat red fish or a toy balloon ready to burst. Even in the herbarium specimens the broadly rounded floral bract of *Vriesea incurvata* is quite different from the sharp-pointed one of *Vriesea inflata.*

Vriesea inflata (Wawra) Wawra, *Itinera Principum S. Coburgi,* p. 161. 1883. *Vriesea carinata* Wawra var. *inflata* Wawra in *Oesterreichische Botanische Zeitschrift,* volume 30, p. 183. 1880. LEAVES many in a broadly funnelform rosette, 2-4 dm. long, usually exceeding the inflorescence, obscurely vestite with minute appressed scales; sheaths broadly elliptic, much broader than the blades, pale brown toward base, usually purplish; blades ligulate, rounded or broadly acute and apiculate, 2-3 cm. wide, flat, green. SCAPE short, straight or somewhat curved, 4-5 mm. in diameter; scape-bracts erect, short but imbricate, broadly elliptic, inflated, apiculate or with a small narrow blade. INFLORESCENCE simple, very dense, elliptic or lanceolate, thick, many-flowered, 11-16 cm. long, 5-7 cm. wide, glabrous. FLORAL BRACTS all densely imbricate, covering the sepals, very broadly ovate, acute, carinate, inflated, 4-5 cm. long, red with yellow or green margins; pedicels 8 mm. long. FLOWERS erect with the petals all issuing from one side of the inflorescence. SEPALS elliptic, obtuse, 25-35 mm. long, ecarinate, nerved. PETALS ligulate, obtuse, yellow or toward apex greenish, to 45 mm. long, bearing two large scales at base. STAMENS and STYLE exserted.

13. Vriesea carinata

Wawra was much more impressed by his *Vriesea carinata* than by his *Vriesea inflata,* and he illustrated it in color in his report of the expedition with Emperor Maximilian. In the century since its discovery it has equally impressed other botanists and horticulturists with its small slender form and bright colors and it has been popular in cultivation not only for its own image but also as the parent of a great number of hybrids.

In a group of species that is notable for its hybrids, *Vriesea carinata* stands out for its ability to cross in all directions and leave its undeniable stamp on the progeny. It not only responds to man's efforts with alacrity but frequently hybridizes in the wild, thereby puzzling the systematic botanist no end.

Its hybrid, *Vriesea morreniana,* is a case in point. This cross was produced artificially in Eduard Morren's gardens and illustrated between its parents, *Vriesea carinata* and *Vriesea psittacina.* However, *Vriesea morreniana* or its exact replica has been collected in the wild on numerous occasions and sometimes in areas where *Vriesea psittacina* is unknown. This apparent contradiction is easy to understand, as the only effect of the second parent is to lengthen and loosen the spike, the floral bracts and flowers remaining practically indistinguishable from those of *Vriesea carinata.*

Vriesea carinata is common in the coastal rainforest from Espirito Santo to Santa Catarina, and consequently new stocks of live plants are constantly being added to gardens everywhere. Margaret Mee's plant is another from Caraguatatuba, which must be a bromeliad Eden. It is a thoroughly typical example of the species and closely resembles Wawra's original.

Vriesea carinata Wawra in *Oesterreichische Botanische Zeitschrift,* volume 12, p. 349, 1862. PLANT typically flowering 24-36 cm. high, propagating by very short erect shoots. LEAVES some 12-24 in a funnelform rosette, 15-27 cm. long, so sparsely and minutely lepidote as to appear glabrous; sheaths broadly elliptic, distinct, pale green and usually tinged with red; blades ligulate, broadly acute or sometimes rounded and apiculate, 14-20 (-32) mm. wide, flat, thin, pale green. SCAPE erect or ascending, very slender; scape-bracts erect and closely enfolding the scape, slightly longer or shorter than the internodes, elliptic, apiculate, membranaceous, pale green or purplish or reddish. INFLORESCENCE usually exceeding the leaves, simple, dense or subdense, few-flowered, nearly as broad as long, oblong with a triangular-retuse apex, strongly compressed, glabrous; rhachis geniculate, slender, the internodes very short. FLORAL BRACTS many times longer than the internodes but spreading and so narrow that they generally do not cover the rhachis, subtriangular, 30-36 (-44) mm. long, usually exceeding the sepals, 12-16 mm. wide, sharply carinate, strongly incurved toward apex, membranaceous, apex and margins yellow or green and the remainder usually bright red. FLOWERS subsessile. SEPALS sublanceolate, rounded and apiculate, *ca.* 30 mm. long, carinate. PETALS linear, obtuse, 35-45 (-60) mm. long, yellow except the green apex, bearing two large obtuse scales at base. STAMENS and STYLE exserted. CAPSULE shorter than the sepals.

14. Vriesea ensiformis variety bicolor

Typical *Vriesea ensiformis* was discovered by Vellozo way back around 1790 and its single lax spike is a familiar feature of the coastal rain-forest from Bahia to Santa Catarina. Its variety *bicolor,* on the other hand, was not discovered until more than a century later and still is not known outside of its original state of São Paulo.

Variety *bicolor* has floral bracts that are red at base and yellow toward apex compared to wholly red ones in the typical variety. This bicolorous character is not all unusual in *Vriesea* and by itself would be scarcely worth noting, but it is accompanied by a great softening in the texture of the upper half of the floral bract, rapidly followed by a blackening of the extreme apex as if charred.

This apparent deterioration is rather rare in the floral bracts of *Vriesea* but is not uncommon in leaf-tips. Evidently no study has been made of the phenomenon but its regularity would argue that it is, at least initially, an integral part of the plant's physiology and in no wise pathological. Whatever the cause of the coloration of variety *bicolor,* the net result is highly ornamental, as Margaret Mee's plant from Paranapiacaba ably demonstrates.

Vriesea ensiformis (Vellozo) Beer variety *bicolor* L. B. Smith in *Arquivos de Botânica do Estado de S. Paulo,* nova série, volume 1, p. 116. 1943. PLANT flowering 4-14 dm. high but usually less than 1 m. LEAVES 20-30 in a broadly funnelform rosette, 3-9 dm. long, very finely and obscurely lepidote; sheaths elliptic, distinct, many times shorter than the blades; blades ligulate, broadly acute or rounded and apiculate, slightly narrowed at base, 3-5 cm. wide, green. SCAPE erect, stout; scape-bracts erect, imbricate, elliptic, apiculate. INFLORESCENCE simple, many-flowered, 2-7 dm. long, glabrous; rhachis stout, straight or nearly so. FLORAL BRACTS spreading with the flowers at anthesis and exposing the rhachis, usually equalling or exceeding the sepals and 3 times the internodes or more, broadly subelliptic, usually with inrolled margins enfolding the sepals, to 55 mm. long, smooth or obscurely nerved on drying, in the variety *bicolor* with red coriaceous bases and soft yellow apices that soon blacken and disintegrate. FLOWERS never secund; pedicels short, stout. SEPALS elliptic, thick, 30-40 mm. long, yellow. PETALS linear, to 5 cm. long, yellow, bearing two large scales at base. STAMENS exserted. CAPSULE about equaling the sepals.

15. Vriesea guttata

Seed of *Vriesea guttata* was brought to Europe from Santa Catarina in 1870, and by 1875 had developed its handsomely dark-spotted leaves to such a degree that Linden and André could wait no longer for the flowers and described it as a new species on the basis of leaves alone, guessing correctly that it was a *Vriesea*. Three years later it began to flower in several European gardens and in 1880 Eduard Morren completed its description in the "Belgique Horticole" and illustrated it with a large colored plate. Ten years is a long time to wait between seed and flower but it does suggest an alternative in the importation of tender species that cannot survive fumigation in the growing stage.

Vriesea guttata ranges north to the Organ Mountains near Rio, where Margaret Mee collected the subject of this plate; but its true home and greatest development would seem to be in Santa Catarina where originally collected. There Padre Raulino Rietz, our present Brazilian authority on bromeliads, has found it in at least six different localities in the coastal rainforest while it appears rather rare elsewhere.

Vriesea guttata Linden & André in *L'Illustration Horticole,* volume 22, p. 43, plate 200. 1875 (leaves only) ; E. Morren in *Belgique Horticole,* volume 30, p. 13, plate 1-3. 1880 (complete). PLANT flowering 4-7.5 dm. high with the inflorescence extended. LEAVES less than 20 in a funnelform rosette, 15-40 cm. long, densely and minutely appressed-lepidote on both sides, marked with red-brown spots; sheaths elliptic, slightly wider than the blades, more or less castaneous; blades ligulate, broadly rounded and apiculate, flat, 25-40 mm. wide. SCAPE decurved, slender; scape-bracts slightly exceeding the internodes, broadly elliptic, apiculate. INFLORESCENCE simple, many-flowered, oblong, acute, strongly complanate, 10-35 cm. long, 4-6 cm. wide; rhachis flexuous, angled. FLORAL BRACTS spreading at anthesis or at least the basal ones, broadly elliptic, acute, 3-4 cm. long, 3-4 times as long as the internodes, 26 mm. wide, obtusely carinate toward apex, never incurved, subcoriaceous, rose, farinose, smooth. FLOWERS never secund; pedicels obconic, stout. SEPALS narrowly elliptic, acute, to 3 cm. long, usually equalling or exceeding the bracts, farinose. PETALS ligulate, very narrow, obtuse, 35-45 mm. long, yellow, bearing two entire scales. STAMENS exserted.

16. Vriesea simplex

Vellozo really missed an opportunity when he gave this curious bromeliad such a humdrum species name as "simplex," but Eduard Morren in naming its only near relative was struck by the primitive ladderlike outline of the inflorescence and named his species "scalaris." The two species are very distinct from all other Vrieseas but confusingly close to each other. However, Vellozo's plate shows quite unmistakably the more robust form of *Vriesea simplex* as well as its much ampler floral bracts, so there can be little doubt of the accuracy of Morren's distinctions in separating the more slender and graceful *Vriesea scalaris*.

For a species that has been known so long and so widely—from eastern Brazil to Colombia—*Vriesea simplex* is rather rare. Quite possibly it is one more example of a species that is so successful at vegetative propagation that it has no great need of flowers. In that case we are fortunate in having available a technique that will force reluctant bromeliads into flower.

Margaret Mee's plant came from the grounds of the Instituto de Botânica at Ipiranga, birthplace of Brazilian independence, now swallowed up in the bulging city of São Paulo. It could have been native, as there have been collections from the wild nearby; or it could have been cultivated in a land where bromeliad cultivation involves only prying loose from one tree and tying onto another.

Vriesea simplex (Vellozo) Beer, *Die Familie der Bromeliaceen,* p. 97. 1857. *Tillandsia simplex* Vellozo, *Flora Fluminensis,* p. 133. 1825, and *Icones, volume* 3, plate 130. 1835. PLANT flowering 4–7 dm. high with the inflorescence extended. LEAVES 10-15 in a funnelform rosette, 3–4 dm. long; sheaths broadly elliptic, 7–10 cm. long, minutely lepidote; blades linear-oblanceolate, broadly acute or rounded and apiculate, 15–35 mm. wide, flat, obscurely lepidote beneath, concolorous or reddish beneath. SCAPE decurved, very slender; scape-bracts erect, broadly elliptic, apiculate, chartaceous, red with yellow or green apices. INFLORESCENCE simple, very lax, 2-3 dm. long, 8-12 flowered, glabrous; rhachis, flexuous to strongly geniculate. FLORAL BRACTS spreading, broadly ovate, 35-50 mm. long and more than half as wide, completely enfolding the base of the flower and its margins overlapping, carinate toward apex, chartaceous. FLOWERS not secund; pedicels slender, 8-20 mm. long. SEPALS elliptic, obtuse, 3–4 cm. long, usually much exceeding the bracts. PETALS 10-15 mm. longer than the sepals, yellow with blue-green apices. STAMENS exserted.

17. Neoregelia hoehneana

With the genus *Neoregelia* we come to the third and last subfamily of the *Bromeliaceae,* the *Bromelioideae.* All the members of this subfamily have an inferior ovary and a fruit that is technically a berry, but frequently a very tough, dry and leathery one. Most of them have spiny leaves as well.

Neoregelia itself has compact inflorescences sunk in the center of the rosette and flowers that are pedicellate. Several other genera have such nidular or "nested" inflorescences or have the pedicellate flowers—but never the two together. In addition most neoregelias have spreading violet to white petal-blades, but here we are entering territory in dispute with *Nidularium.*

The majority of neoregelias are esteemed in cultivation because of their bright colors, but *Neoregelia hoehneana* appeals almost wholly through its graceful and curious form. The slender rhizomes arch away from all support and produce tiny rosettes that are truly air plants living in total suspension. The small white flowers are hidden until the artist tilts the plant in our direction.

Like most of its kind, *Neoregelia hoehneana* has a very limited distribution. Margaret Mee found her present subject at Caraguatatuba, where the original was collected in 1940—it has never been found elsewhere.

The species was named in honor of Dr. Frederico Carlos Hoehne, former director of the Instituto de Botânica and founder of the Flora Brasilica for which all of these plates were originally prepared.

Neoregelia L. B. Smith in *Contributions from the Gray Herbarium of Harvard University,* volume 104, p. 78. 1934. Perennial herbs propagating by lateral branches or elongate rhizomes. Leaves rosulate, coarsely spinose-serrate to subentire. Scape short and wholly concealed by the leaf-sheaths. Inflorescence simple or rarely compound but always short, very dense and sunk in the center of the rosette. Flowers pedicellate. Sepals asymmetric, more or less connate. Petals unappendaged, connate; blades spreading, usually acute or attenuate, violet, blue or white. Stamens and pistil included.

Neoregelia hoehneana L. B. Smith, *Bromeliaceae of Brazil in Smithsonian Miscellaneous Collections,* volume 126, p. 28, 150, fig. 56. 1955. PLANT propagating by slender arching-pendent rhizomes. LEAVES about 10 in each rosette, even the inner ones only 15 cm. long, bearing minute dark appressed scales; sheaths ample, 6 cm. long, forming a subcylindric or ellipsoid tank, densely lepidote; blades linear, flat, broadly rounded and apiculate, 25 mm. wide, green, very sparsely lepidote, very sparsely serrulate with teeth less than 0.5 mm. long. INFLORESCENCE few-flowered, hidden in the center of the rosette. FLORAL BRACTS oblong, obtuse, about equalling the sepals, thin, greenish white, subglabrous. PEDICELS slender, 5 mm. long. SEPALS short-connate, 12 mm. long, broadly obtuse, strongly asymmetric, green medianly. PETALS 22 mm. long, white; blades elliptic, obtuse or apiculate. OVARY ellipsoid, 7 mm. long.

18. Nidularium burchellii

The genus *Nidularium* originally included the species that now compose *Neoregelia* and the present arbitrary separation of the two genera on whether the flowers are sessile or pedicellate is not completely satisfactory by any means. It leaves *Neoregelia* smaller and more clearly defined, but places *Nidularium burchellii* and *N. microps* in *Nidularium* to the complication and confusion of that genus.

Nidularium burchellii has flat spreading petal-blades like those of a *Neoregelia,* but in addition to its sessile flowers it has a long enough scape to lift the inflorescence above the leaf-sheaths as in many nidulariums. Also the upper scape-bracts that make an involucre about the inflorescence are much more like those of typical *Nidularium.*

The species is named in honor of W. J. Burchell, an English botanist who crossed Brazil from Santos in São Paulo to Belém in Pará. Travelling by pack train and dugout canoe, it took him the better part of five years, but he made a remarkable collection in areas seldom or never revisited (see L. B. & R. C. Smith, Itinerary of William John Burchell in Brazil, 1825-1830, in Phytologia, volume 14, p. 492. 1967). He discovered his new *Nidularium* in the coastal rain-forest of São Paulo near Santos and not very far from Paranapiacaba where Margaret Mee collected the one shown here.

Nidularium Lemaire in *Le Jardin Fleuriste,* volume 4, miscellanea, p. 60. 1854. Perennial stemless herbs propagating by lateral shoots and elongate rhizomes. Leaves rosulate, usually serrulate. Scape varying from short and hidden to elongate and much exceeding the leaf-sheaths. Inflorescence always compound with the flowers in flat digitate fascicles in the axils of large ample bracts, usually short and compact, rarely elongate. Floral bracts well developed. Flowers sessile. Sepals mostly subsymmetric, more or less connate. Petals unappendaged, usually high-connate; blades typically erect and cucullate, rarely spreading and flat. Stamens included, the second series highly adnate to the petals. Pistil included; ovules numerous, scarcely appendaged.

Nidularium burchellii (Baker) Mez in DeCandolle, *Monographiae Phanerogamarum,* volume 9, p. 101. 1896. *Aechmea burchellii* Baker in *Journal of Botany,* volume 17, p. 231. 1879. PLANT propagating by long slender horizontal rhizomes. LEAVES few in a subcylindric rosette, sometimes short and broad, sometimes long and narrow, 2-5 dm. long; sheaths broadly elliptic, entire or serrate toward apex, diffusely vestite with brown subappressed scales; blades ligulate, more or less narrowed toward base, attenuate or acute and apiculate, 2-5 cm. wide, glabrous or obscurely lepidote, wholly green or tinged beneath with red, narrowly channeled, subdensely serrulate with spines 0.5-1 mm. long. SCAPE short and about equalling the leaf-sheaths, ferruginous-furfuraceous; lower scape-bracts lanceolate, acuminate, minutely serrulate, the upper (involucral) shorter than the inflorescence, ovate, acute, serrulate, green. INFLORESCENCE densely subglobose, 25-55 mm. long, densely ferruginous-lanate; fascicles few-flowered. FLORAL BRACTS ovate, acute, equalling the sepals or slightly shorter, strongly serrate, recurved near apex, green. SEPALS connate for 2 mm., 10 mm. long, subsymmetric, broadly elliptic, acute. PETALS 18 mm. long, connate for 1-2 mm., white; blades spreading, flat, elliptic, acute. STAMENS included; filaments of the second series highly adnate to the petals; pollen 4-pored. OVARY ellipsoid, 7 mm. long; epigynous tube large, funnelform; placentae apical; ovules obtuse.

19. Nidularium billbergioides

Except for its unusually long and nearly bare scape, *Nidularium billbergioides* is quite typical of its genus, as evidenced by its large primary bracts and erect petal-blades. The same long scape was evidently the reason for the specific name that likened it to a billbergia, because the only possible candidate at that time would have been *Billbergia pyramidalis,* a rather far-fetched comparison otherwise. However, if it were not for a ruling of the International Code that governs our use of botanical names, we might not be using "billbergioides" at all, but instead "terminale" from a much earlier discovery and publication by Vellozo.

A further confusion in names arises from the unusual dominance of yellow in the primary bracts. While the bracts of most nidulariums are distinctly on the red side, these always have a distinct admixture of yellow and usually are nearly pure yellow. Struck by the unusual color in live specimens, Mez published a *Nidularium citrinum,* which Padre Reitz later reduced to a variety of *Nidularium billbergioides;* but we have no evidence whatever of the color of the original collection of the species. In such a case there seems to be little point in dividing the species at all.

Martius collected the type of *Nidularium billbergioides* in southern Bahia, Vellozo found the species near Rio, Margaret Mee found it at Caraguatatuba in São Paulo; and it extends further south into Santa Catarina. All of its stations are in the coastal rainforest.

Nidularium billbergioides (Schultes filius) L. B. Smith in *Contributions from the Gray Herbarium of Harvard University,* volume 95, p. 42. 1931. *Hohenbergia billbergioides* Schultes filius in Roemer and Schultes, *Systema Vegetabilium,* volume 7, part 2, p. 1253. 1830. PLANT propagating by long ascending rhizomes. LEAVES 10-16 in a dense funnelform rosette, 3-7 dm. long, usually much exceeding the inflorescence, wholly green; sheaths broadly elliptic, *ca.* 10 cm. long, entire, covered with minute appressed brown scales; blades ligulate, acute or attenuate, more or less narrowed toward base, 18-38 mm. wide, laxly serrulate, minutely and obscurely lepidote on both sides. SCAPE erect, slender, glabrous; scape-bracts only one or two and usually covering very little of the scape, elliptic, acute. INFLORESCENCE few-flowered but compound; primary bracts broadly ovate to suborbicular with a short or long lanceolate blade, to 7 cm. long, much exceeding the flowers, minutely serrulate, yellow or orange or sometimes brownish red, minutely lepidote, usually recurved toward apex. FLORAL BRACTS very broadly ovate, acute, entire, slightly exceeded by the sepals, membranaceous, sparsely lepidote. FLOWERS 25-28 mm. long. SEPALS connate for 5 mm., slightly asymmetric, elliptic, apiculate, 12-15 mm. long. PETALS connate for two-thirds of their length, broadly rounded and minutely apiculate, white. OVARY obovoid; epigynous tube short: placentae subapical; ovules obtuse.

20. Nidularium seidelii

Its elongate floral axis immediately distinguishes *Nidularium seidelii* from all other nidulariums, but except for that and its yellow coloration, it is a completely orthodox member of the genus with flat hand-like fascicles of flowers in the axils of large, ample bracts. At one time *Nidularium loeseneri* was thought to be a similar species of *Nidularium* with an elongate floral axis, but it now develops that this is a bigeneric hybrid that owes its long axis to its second parent, *Aechmea calyculata.*

Nidularium seidelii is named for Alvim Seidel of Corupá, Santa Catarina, a horticulturist who is particularly interested in bromeliads and who has been instrumental in introducing a number of unusual species into the United States. Although it is not certain who was the first to discover it at Ubatuba on the coast of São Paulo, Mr. Seidel was the first to bring this new nidularium to public attention.

Nidularium seidelii L. B. Smith & R. Reitz in *Phytologia,* volume 9, p. 257, plate 4, figs. 11-13. 1963. PLANT flowering 35-45 cm. high. LEAVES rather many in a broadly funnelform rosette, 5-8 dm. long, much exceeding the inflorescence; sheaths elliptic, 12-15 cm. long, entire, densely vestite with appressed brown-centered cinereous scales; blades ligulate, broadly subacute and apiculate, slightly narrowed at base, to 4 cm. wide, obscurely lepidote. SCAPE erect, 6 mm. in diameter, nearly naked, finely ferruginous-lanate becoming more or less glabrous; scape-bracts 2, remote, the upper broadly ovate, 35-70 mm. long, serrulate, lepidote. INFLORESCENCE subcylindric, 11-15 cm. long, 5-8 cm. wide; primary bracts suberect and densely imbricate, like the upper scape-bract but larger, much exceeding and concealing the flowers, yellow; branches pulvinate, few-flowered. FLORAL BRACTS elliptic, acute, about equalling the sepals, thin, entire, sparsely lepidote. FLOWERS sessile, pale-ferruginous-lanate toward base. SEPALS broadly spatulate, truncate and mucronulate, slightly asymmetric, 14 mm. long, connate for 2.5 mm., prominently nerved. PETALS connate by their claws, naked; blades elliptic, obtuse, 10 mm. long, white. OVARY ellipsoid; epigynous tube broadly funnelform, 1.5 mm. high; placentae apical.

21. Wittrockia superba

In general appearance, *Wittrockia* seems closely allied to *Neoregelia* and *Nidularium,* with its dense cluster of flowers sunk in the center of the rosette; but close examination reveals a pair of scales at the base of each petal and a most unusual type of fusion, with the petals free at base but joined above, leaving triangular peepholes showing the bases of the alternate filaments.

Wittrockia is one of the many marks of Swedish research in Brazil. In this case, Regnell came to Brazil from Sweden in 1840 and after much pioneer work in southern Minas set up a fund for its continuation. His successor, Mosén, collected *Wittrockia superba* near Santos, but it was finally described by Lindman, who named it in honor of his former teacher, Veit Brecher Wittrock, distinguished more in the botany of the Old World than of Brazil.

Fortunately for us, Mosén was an excellent collector and preserved flowers in alcohol and made excellent pencil sketches as well. Thus Lindman was able to publish a very good account of *Wittrockia superba,* so good that only a color plate could improve on it and show just how "superb" it really is. The handful of later species include some beauties, but none quite its equal.

Margaret Mee found her plant at Caraguatatuba and the species extends further north to Rio and south to Santa Catarina in the coastal rainforest.

Wittrockia Lindman in *Kongl. Svenska Vetenskaps-Akademiens Handlingar,* volume 24, no. 8, p. 15, 20. 1891. Perennial herbs propagating by rhizomes. Leaves rosulate, serrate or serrulate. Scape short and hidden to elongate. Inflorescence compound with very short branches, densely subglobose or hemispheric; primary bracts large, ample. Floral bracts well developed. Flowers subsessile. Sepals free or somewhat connate, more or less asymmetric. Petals free at base, connate or agglutinated above base, appendaged. Stamens included; pollen biporate. Ovary inferior; placentae subcentral; ovules obtuse. Fruit scarcely enlarged from the ovary.

Wittrockia superba Lindman in *Kongl. Svenska Vetenskaps-Akademiens Handlingar,* volume 24, no. 8, p. 20, plate 2, figs. 13-21. 1891. LEAVES many in a dense rosette, 8-10 dm. long; sheaths large, very broadly obovate, very densely brown-lepidote; blades ligulate, acute, linear from the base to the middle and then enlarged between the middle and apex, 5-7 cm. wide, subdensely serrate with spines to 4 mm. long, green, red at apex. SCAPE short, stout; scape-bracts imbricate, ovate, attenuate, the lowest spinose-serrate, the upper entire, massed below the inflorescence and equaling it, red. INFLORESCENCE densely bipinnate, many-flowered, broadly subhemispheric; primary bracts like the upper scape-bracts but narrower;

branches with naked sterile bases 3-4 cm. long and flowers fascicled at the apex. FLORAL BRACTS lance-ovate, attenuate, exceeding the sepals, membranaceous but much thickened toward apex, pungent, entire. SEPALS strongly asymmetric, ovate, attenuate, pungent, 23 mm. long, connate for 5 mm. PETALS shorter than the sepals, acute, connate for 5-6 mm., blue toward apex, bearing two fimbriate scales at base. STAMENS included; pollen ellipsoid, reticulate, biporate. OVARY subcylindric, 15-20 mm. long; epigynous tube large; placentae small, central; ovules obtuse.

22. Hohenbergia horrida

Of all bromeliad genera, *Hohenbergia* with its numerous small cones is one of the easiest to recognize. Of course there are some technical details like the appendaged petals and the nearly complete absence of an epigynous tube that distinguish it from an occasional *Gravisia* or *Streptocalyx,* but really these distinctions are scarcely necessary. *Hohenbergia* has a very distinctive distribution also, with 16 species in eastern Brazil, 15 in tiny Jamaica, and very few anywhere else.

One glance at this spiniest of hohenbergias shows that its name of "horrida" is well-earned and leads to the correct guess that it comes from arid territory. In this case, Margaret Mee collected it at Brejo de Madre de Dios in eastern Pernambuco, where thorn-scrub often alternates with bare hot sand that burns even through boot soles. Like *Encholirium spectabile* from the same region, it is a prime candidate for the cactus garden.

Hohenbergia Schultes filius in Roemer & Schultes, *Systema Vegetabilium,* volume 7, part 2, p. lxxi, 1251. 1830. Perennial often epiphytic herbs. Leaves rosulate; blades ligulate, spinose. Scape erect, always evident. Inflorescence bipinnate or tripinnate, composed of dense cone-like spikes. Floral bracts well developed. Flowers perfect, sessile, compressed. Sepals nearly or quite free, distinctly asymmetric, mucronate. Petals nearly or quite free, bearing two scales on the inner surface; blades acute, suberect to spreading at anthesis. Stamens shorter than the petals; filaments of the second series highly adnate to the petals; pollen with two or four pores. Ovary wholly inferior, compressed and subalate, changing little in fruit; epigynous tube short or none; ovules several in each cell, obtuse or caudate.

Hohenbergia horrida Harms in *Notizblatt,* volume 12, p. 525. 1935. PLANT flowering 1 m. high. LEAVES in a funnelform rosette, 6-8 dm. or longer, rigid; sheaths very broad, lepidote; blades ligulate, attenuate to a stout spine, nerved, laxly serrate with antrorse and retrorse spines 5 mm. long. SCAPE erect, elongate, sparsely lanate; scape-bracts lanceolate, pungent, 6-10 cm. long, entire. INFLORESCENCE paniculate, many-flowered, 5 dm. long, tripinnate and lax at base, bipinnate and denser toward apex; axes lanate; primary bracts lanceolate, attenuate, 7-11 cm. long; lower branches to 13 cm. long; secondary bracts lanceolate, usually shorter than the spikes; spikes many, sessile or subsessile, 3-4 cm. long. FLORAL BRACTS broadly ovate, 12-18 mm. long, tapering to a pungent spine 4-6 mm. long, 10 mm. wide, rigid, strongly nerved, sparsely lanate, soon nearly glabrous. SEPALS strongly asymmetric with a wing 4 mm. wide, to 10 mm. long with a terminal spine 3-3.5 mm. long, carinate. PETALS to 20 mm. long, purple. OVARY 4 mm. long, villous.

23. Aechmea orlandiana

Almost every botanist has a different idea of what belongs in *Aechmea.* Some are impressed by differences and separate *Chevalieria, Lamprococcus, Hoplophytum, Ortgiesia, Pothuava* and *Macrochordion* as genera, while others influenced by similarities and intergradations consider these no better than subgenera at the most. It is easy to find fault with either point of view, much more difficult to make a usable system. In reality, *Aechmea* is a large group of species developing in all directions at once from a starting point that is a rosulate plant with a panicle of perfect flowers, well-developed floral bracts, mucronate sepals and appendaged petals. From there on it is like English grammar with more exceptions than rules, but always a modicum of regularity somewhere to tie the errant species to the central core. *Aechmea* is a large genus of over 150 species with a large range from Mexico and the West Indies to Argentina.

In the case of *Aechmea orlandiana,* Margaret Mee had to use material cultivated in the Botanic Garden in Rio instead of collecting her own. The original stock was found at Itapemirim in Espirito Santo along with a number of other prizes by Mulford B. Foster, all-time record holder for the discovery of new and beautiful bromeliads. At his request is was named in honor of his home town of Orlando, Florida, and was specially featured in its anniversary celebrations.

However, it is not necessary to beat any drums to rouse interest in a species that is so highly ornamental in both leaves and flowers. Also the species is recognizable by the leaves alone because of their peculiar embossed effect of raised dark and sunken lighter green bands.

Aechmea Ruiz & Pavon, *Florae Peruvianae et Chilensis Prodromus,* p. 47. 1794. Plants perennial, large or medium-sized, often epiphytic. Leaves rosulate; blades usually ligulate. Scape conspicuous in most species. Inflorescence compound, or simple, very variable. Floral bracts conspicuous to nearly absent. Flowers usually perfect, usually sessile. Sepals mostly asymmetric with a large wing on the right, usually free or nearly so. Petals free, bearing two scales. Stamens not forming a tube; filaments of the second series more or less adnate to the petals; pollen mostly with two or four pores. Ovary inferior; ovules caudate or obtuse. Berry usually dry; seeds small, rugose, dark.

Aechmea orlandiana L. B. Smith in *Arquivos de Botânica do Estado de S. Paulo,* nova série, volume 1, fasciculo 3, p. 55, plate 69. 1941. PLANT propagating by short ascending shoots, flowering over 4 dm. high. LEAVES about 20 in a dense funnelform rosette, to 3 dm. long, marked beneath with broad dark spots or bands; sheaths large, very broadly elliptic, entire, dark purple beneath, covered on both sides with minute appressed brown scales; blades ligulate, broadly acute and apiculate, to 45 mm. wide, obscurely punctulate-lepidote, serrate with straight spines 4 mm. long. SCAPE erect, slender, red, glabrous; scape-bracts erect and enfolding the scape, broadly elliptic, apiculate, serrulate toward apex, membranaceous, red, minutely white-flocculose, the lower shorter than the internodes, the upper imbricate. INFLORESCENCE very densely bipinnate, ovoid, 7 cm. long, 5 cm. in diameter; primary bracts like the scape-bracts, also serrulate, exceeding the spikes; spikes subsessile, dense, distichously 4-flowered, *ca.* 3 cm. long. FLORAL BRACTS broadly ovate, acuminate, entire, exceeding the sepals, red, the margins free from the rhachis. FLOWERS sessile, erect. SEPALS free, strongly asymmetric, suboblong, short-mucronulate, 11 mm. long, nerved, glabrous. PETALS 2 cm. long; blades erect, elliptic, broadly rounded, pale yellow except the white margins, bearing fimbriate scales at base. STAMENS included; pollen biporate. OVARY slenderly ellipsoid, 6 mm. long; epigynous tube short; placentae central; ovules obtuse.

24. Aechmea fosteriana

In many ways *Aechmea fosteriana* is a sequel to *Aechmea orlandiana.* It too was collected by Mulford Foster in Espirito Santo, but three days later and at Vitoria. It has beautiful leaves and flowers, but its colors are softer. The leaves have pastel tones quite different from those of *Aechmea orlandiana,* and the much more open inflorescence has bracts of darker duller red.

Perhaps it is not wholly coincidence that our two outstanding examples of variegated leaves were collected by the same person. The design that is so striking in a room or a glasshouse is the most effective camouflage in the checkered light of the rainforest and takes a very keen eye to detect.

Again Margaret Mee resorted to cultivated material, this time in the gardens of Rio's famed landscape architect, Dr. Roberto Burle Marx; but in so doing she has given us an accurate and highly pleasing record of the coloration of this unusual species.

Aechmea fosteriana L. B. Smith in *Arquivos de Botânica do Estado de S. Paulo,* nova série, volume 1, fasciculo 3, p. 54, plate 67. 1941.

PLANT flowering 6-9 dm. high. LEAVES cylindric-rosulate, over 6 dm. long, densely punctulate-lepidote on both sides; sheaths elliptic, large, dark castaneous; blades erect or slightly divergent, ligulate, 6-10 cm. wide, subdensely serrate with straight brown teeth 3 mm. long, bearing brown-purple spots and a large dark purple one at apex. SCAPE erect, slender, glabrous, dark purple; scape-bracts much shorter than the internodes, elliptic, acute, membranaceous, punctulate-lepidote. INFLORESCENCE paniculate, laxly 3-pinnate; primary bracts like the scape-bracts, exceeding the branches, dark red within; spikes laxly 6-flowered, glabrous; rhachis geniculate. FLORAL BRACTS suborbicular, apiculate, to 13 mm. long, exceeding the ovary, broadly convex, ecarinate, red. FLOWERS sessile, divergent, slightly more than 2-ranked. SEPALS oblong, strongly asymmetric, broadly rounded and apiculate, 10 mm. long, green. PETALS 16 mm. long, yellow, bearing two fimbriate scales at base; blades broadly elliptic, subacute. STAMENS included; filaments of the second series highly adnate to the petals; pollen biporate. OVARY turbinate; epigynous tube crateriform, 3 mm. long; placentae apical; ovules apiculate.

25. Aechmea caudata

Aechmea caudata and its near relatives constitute the subgenus, or if you will, genus *Hoplophytum* distinguished by small sepals fused together for about half their length and then terminating in a long spine. The form of the flowers is much the same in all species, but the inflorescence varies from dense simple spikes to the pyramidal panicle of *Aechmea caudata*. The petals may be white, yellow or blue and sometimes this is the only sharp distinction.

In recent years Padre Raulino Reitz has made a special study of this group, which reaches its maximum of complexity and perplexity in his home state of Santa Catarina. Among other things he has found that while some species commence blooming at the base in normal fashion like *Aechmea,* others begin in the middle and proceed in both directions.

There is no record of the origin of Margaret Mee's plant that was in cultivation in the gardens of the Instituto de Botânica in the city of São Paulo, but in all probability its forbears came from the vicinity of Santos where Mosén collected the type on the Ponta de Taipú in 1875. *Aechmea caudata,* a plant of rocky shores, has been pushed out of some of its old haunts by the growth of Santos, but there need be no concern about its survival since it extends from Espirito Santo to Santa Catarina.

Aechmea caudata Lindman in *Kongl. Svenska Vetenskaps-Akadamiens Handlingar,* volume 24, No. 8, p. 29, plate 6, figs. 1-9. Feb. 1891. PLANT flowering to 9 dm. high. LEAVES 10-15 in a broadly funnelform rosette, 5-10 dm. long, minutely appressed-lepidote; sheaths ovate or elliptic, distinct or obscure, entire, brown toward base; blades ligulate, broadly rounded and apiculate, to 8 cm. wide but often much narrower, laxly serrulate with spines less than 1 mm. long, sometimes dark toward apex. SCAPE usually erect, 4-10 mm. in diameter, white-flocculose; scape-bracts erect, lanceolate, acute, entire, membranaceous, red, subglabrous, the upper usually much longer than the internodes. INFORESCENCE densely or subdensely paniculate, 10-25 cm. long, to 11 cm. in diameter, bipinnate to the middle or higher and the remainder simple, white-flocculose; primary bracts like the scape-bracts, usually shorter than the branches; spikes spreading, laxly 4-7 flowered; rhachis slender, strongly geniculate. FLORAL BRACTS ovate, attenuate to a slender brown spine, 7-17 mm. long, entire, nerved, red, the margins free from the rhachis. FLOWERS sessile, spreading, 18-25 mm. long. SEPALS 7-11 mm. long including the long terminal spine, connate. PETALS ligulate, obtuse, 12-15 mm. long, yellow, turning purplish on drying, bearing two small scales at base. STAMENS shorter than the petals; pollen biporate. OVARY subcylindric; epigynous tube evident, crateriform; placentae extending nearly the full height of the cell; ovules obtuse.

26. Aechmea cylindrata

Aechmea cylindrata is so beautiful that maybe we should just appreciate its ornamental qualities and not explore its tangled history. However, for those who are interested in classification it is a shining—or glaring—example of the mess that is *Aechmea*. In describing the species from a Mosén collection from near Santos, Lindman placed it in *Pothuava* and Mez followed his lead, but in so doing contradicted himself so that it is impossible to reach the species in his key. In reality the sepals of *Aechmea cylindrata* are indistinguishable from those in *Hoplophytum* and if anything are a bit more connate and therefore more characteristic than those of *Aechmea caudata,* which everybody agrees belongs in that subgenus.

Margaret Mee's plant from Miracatu, São Paulo, illustrates another common problem in Brazilian botany—place names. Miracatu does not appear on the American Geographical Society map of 1938 and one might assume that it did not exist then, but comparison with the most recent maps from Brazil show that what was "Prainha" in 1938 is now "Miracatu" on the coastal slope of southeastern São Paulo. Names change frequently and even state capitals are no exception.

Aechmea cylindrata Lindman in *Kongl. Svenska Ventenskaps-Akadamiens Handlingar,* volume 24, No. 8, p. 32, plate 8, figs. 28-35. 1891. PLANT propagating by stout rhizomes 15-20 cm. long, flowering 3-6 dm. high. LEAVES 10-17 in a dense rosette, suberect, 2-8 dm. long; sheaths large but usually not much wider than the blades, elliptic, entire, often tinged with dark purple, covered with minute appressed pale brown-centered scales; blades ligulate, broadly rounded and apiculate, 30-65 mm. wide, pale-lepidote especially beneath, green except for the reddish apex, laxly and minutely serrate or nearly entire. SCAPE erect, white-flocculose at first; scape-bracts narrowly lanceolate, attenuate, entire, pale rose, very delicate and soon deciduous. INFLORESCENCE simple, dense or lax at anthesis, 6-25 cm. long, 25-55 mm. in diameter, sparsely white-flocculose. FLORAL BRACTS narrowly triangular, attenuate, thin with a slightly pungent apex, the lower exceeding the flowers. FLOWERS polystichous, sessile, 14-20 mm. long. SEPALS asymmetric, to 5 mm. long without the brown terminal 2 mm. spine, connate for 3 mm., pale rose. PETALS ligulate, obtuse, to 13 mm. long, pale blue, bearing two fimbriate scales at base. STAMENS included. OVARY slenderly ellipsoid, strongly nerved; epigynous tube large; placentae central; ovules obtuse.

27. Aechmea vanhoutteana

Professional botanists consider it somewhat immodest to name a new species for oneself, but in business it is normal to advertise one's product as personally and as forcefully as possible. Flemish horticulturist Heer Van Houtte was a sound businessman, so when his gardens produced this jeweled scepter of a bromeliad, he did not hesitate nor thumb the dictionaries for an adequately descriptive Latin or Greek word, he simply put his trademark on it and from then on evermore it was "vanhoutteana." Incidentally this business practice continues to be quite popular today.

Van Houtte did not divulge the source of his prize and it is quite possible that he did not know it, but since his time it has been collected on the Serra do Picu in Minas and on Itatiaia, while Margaret Mee's specimen came from the Campos do Jordão in São Paulo. All three localities are high on the mountain chain that divides the coastal rainforest from the westward sloping planalto.

With its simple inflorescence and nearly free short-mucronate sepals *Aechmea vanhoutteana* is a bona fide member of the subgenus *Pothuava,* where *Aechmea cylindrata* had to be excluded. Although there is the additional distinction that the inflorescence is always simple in *Pothuava* and usually compound in *Hoplophytum,* the line between the two seems very thin indeed.

Aechmea vanhoutteana (Van Houtte) Mez in Martius, *Flora Brasiliensis,* volume 3, part 3, p. 366. 1892. *Echinostachys vanhoutteana* Van Houtte, *Van Houtte Catalogue.* 1878. PLANT flowering to 9 dm. high. LEAVES *30-50* in a dense funnelform rosette, suberect, rigid, covered beneath with a pale membrane of fused scales; sheaths broadly ovate, large, tinged with dark purple, pale-lepidote above; blades ligulate, acute or attenuate with a stout terminal spine, 3-7 cm. wide at base, glabrous above, subdensely serrate with stout antrorse brown spines 7 mm. long. SCAPE erect, stout; scape-bracts densely imbricate and covering the scape, large, broadly elliptic, rounded and strongly mucronate, coriaceous, entire or sparsely and minutely serrate, covered beneath with a pale membrane of fused scales. INFLORESCENCE simple, cylindric, obtuse, 7-24 cm. long, 4-5 cm. in diameter, very densely many-flowered, bearing a small setose coma at apex. FLORAL BRACTS many-ranked, suboblong, acute and mucronate, from slightly shorter than the sepals to exceeding them, enfolding the ovary, red toward apex, pale and white-lanate toward base, sparsely dentate, strongly nerved, submembranaceous. FLOWERS sessile, 20-25 mm. long. SEPALS slightly connate, asymmetric, oblong, acute and mucronate, 8-10 mm. long, lanate toward base. PETALS ligulate, obtuse, to 18 mm. long, white except the blue apex, bearing two oblong fimbriate scales at base. STAMENS much shorter than the petals. OVARY subellipsoid, 7 mm. long; epigynous tube large; placentae central; ovules obtuse.

28. Quesnelia humilis

Like a small nation that survives because two jealous neighbors keep each other from annexing it, so *Quesnelia* survives between *Aechmea* and *Billbergia.* If it were not for the ties with *Billbergia, Quesnelia* would be no stronger than a subgenus of *Aechmea,* and vice versa. Thus *Quesnelia* has the large unarmed sepals of *Billbergia* and the regular non-coiled petals and porate pollen of *Aechmea.* In habit also it manages to strike a balance between the two.

As its name implies, *Quesnelia humilis* is short in stature, but it compensates for it in coloration. In form it is intermediate in this intermediate genus, its inflorescence being neither so dense as that of *Quesnelia arvensis* or *Q. testudo* nor so lax as that of *Quesnelia liboniana* or *Q. marmorata.*

Margaret Mee got her plant at the Instituto de Botânica on the outskirts of São Paulo where a large tract of virgin rainforest is carefully preserved. *Quesnelia humilis* has never been found outside the state of São Paulo, but it shows few preferences there, going from near sea level in the vicinity of Santos to the crest of the Serra do Mar or Coast Range at Alto da Serra, and then well inland on the planalto.

Quesnelia Gaudichaud, *Atlas, Voyage autour du monde executé 1836-37 sur la corvette La Bonite,* plate 54. 1842. Mostly stemless often epiphytic herbs of medium stature. Leaves rosulate; blades usually ligulate. Scape conspicuous. Inflorescence compound or simple, variable. Floral bracts mostly conspicuous. Flowers perfect, sessile. Sepals more or less asymmetric, free or short-connate. Petals free, regular, erect or nearly so, bearing two scales at base. Stamens shorter than the petals, the second series highly adnate to the petals; pollen grains ellipsoid, biporate. Ovary inferior; epigynous tube distinct; placentae central to apical; ovules numerous, obtuse.

Quesnelia humilis Mez in Martius, *Flora Brasiliensis,* volume 3, part 3, p. 386. 1892. PLANT propagating by short horizontal rhizomes, flowering 3-4 dm. high. LEAVES few in a tubular rosette, 20-56 cm. long, sometimes exceeding the inflorescence, minutely lepidote on both sides; sheaths narrowly elliptic, usually elongate, often dark purple; blades ligulate, rounded and apiculate, 25-45 mm. wide, laxly and minutely serrulate. SCAPE slender, erect, densely white-lanate; scape-bracts erect, slightly to strongly imbricate, narrowly elliptic, the highest rarely to 8 cm. long, the others much smaller, entire, membranaceous, rose, slightly araneous especially toward base, soon glabrous. INFLORESCENCE simple, few-flowered, densely corymbiform; rhachis very short, densely white-lanate. FLORAL BRACTS entire, membranaceous, rose, dimorphic, the lowest large, like the scape-bracts, about equaling the flowers, the others narrowly triangular, exceeded by the sepals. FLOWERS erect, sessile. SEPALS connate for

2 mm., slightly asymmetric, narrowly lanceolate, obtuse, ca. 2 cm. long, membranaceous, red, glabrous or the apex lanate. PETALS erect, 35 mm. long; claw linear, bearing two fimbriate scales at base; blade elliptic with incurved margins, obtuse, red to red-purple. STAMENS included at anthesis; filaments of the second series highly adnate to the petals; pollen ellipsoid, biporate. OVARY ellipsoid, slenderly costate; epigynous tube large; placentae apical.

29. Quesnelia liboniana

In 1848, *Quesnelia liboniana* was discovered in the Organ Mountains by Libon, a professional collector who promptly shipped it home to his employer, DeJonghe, in Brussels. Libon thought it a great rarity, and DeJonghe, proving very successful in growing it, boasted soon after that it was more common in cultivation than in the wild. In reality, Libon must have struck a poor blooming year for the species, as it is one of the commoner bromeliads in the Organ Mountains and is quite abundant in some localities inside the city limits of Rio.

However, this is far from the worst misconception connected with *Quesnelia liboniana.* Once a collector of herbarium specimens took flowering shoots of this species and thrust them into rosettes of a vriesea, and then pressed them to make a fraudulent but very convincing new species. Even Mez, the great authority on bromeliads, was taken in and published a full-page illustration of this so-called *Quesnelia tillandsioides* (see *The Bromeliad Society Bulletin,* volume 16, p. 4. 1966).

Margaret Mee found her specimen of *Quesnelia liboniana* in the classical area of the Organ Mountains, and it is quite typical of the species, with spiny leaves wholly different from those of a Tillandsia or Vriesea.

Quesnelia liboniana (DeJonghe) Mez in *Botanisches Archiv,* Königsberg, volume 1, p. 66. 1922. *Billbergia liboniana* DeJonghe in *Journal d'Horticulture Pratique de la Belgique,* volume 9, p. 1, plate 1, 1851. PLANT propagating by stout rhizomes. LEAVES few in a slenderly funnelform or subcylindric rosette, exceeding the inflorescence, the outer much reduced, the inner to 8 dm. long; sheaths subelliptic, elongate, distinct, covered with appressed pale brown-centered scales; blades ligulate, acute and long-apiculate, 30-47 mm. wide, serrate with straight or curved spines 0.5-2 mm. long, pale-lepidote and sometimes broadly banded beneath, soon glabrous above. SCAPE central, erect, slender, soon glabrous; scape-bracts erect, mostly imbricate, lanceolate, acute, membranaceous, pale brown, pale-lepidote. INFLORESCENCE simple or sometimes with a 1-2 flowered branch at base; primary bract like the scape-bracts. FLORAL BRACTS membranaceous, the lower narrowly triangular, attenuate, exceeding the ovary, the upper ovate, acute, much shorter. FLOWERS sessile, erect to subspreading. SEPALS very short-connate, asymmetric, sublanceolate, acute, to 23 mm. long, red, glabrous. PETALS erect with their apices slightly if at all recurved at anthesis, to 5 cm. long; claw sublinear with two fimbriate scales; blade elliptic, acute, dark blue. STAMENS distinctly shorter than the petals; filaments of the second series half adnate to the petals; pollen ellipsoid, biporate. OVARY cylindric, glabrous; epigynous tube large; placentae central.

30. Quesnelia marmorata

The checkered history of *Quesnelia marmorata* is probably our best illustration of the uncertainty surrounding the genus. Lemaire described the species originally as a billbergia, and it was misidentified with *Billbergia vittata* by Baker and *Billbergia speciosa* by Wittmack. Lindman correctly placed it in *Quesnelia,* but did not realize that it was already known and so gave it a new specific name. Mez transferred the correct species name into *Aechmea* where it remained until recently when Read finally combined the original species name with the correct genus, *Quesnelia.*

So much for the necessary name juggling—now remember one good note about it all, the appropriate original name, "marmorata," which means "marbled" and refers to the ornamental mottling of the leaves. Note also that although the rosette holds water, it manages to do so with only two rows of leaves to form its tall vase. The inflorescence "needs no encomiums, there she is."

Margaret Mee's plant was collected at Parati, down the coast from Rio, in Vellozo's old hunting grounds. Although the species ranges from Espirito Santo to São Paulo, it has been found most frequently on the various peaks in Rio and has made itself thoroughly at home in the Jardim Botânico.

Quesnelia marmorata (Lemaire) R. W. Read in *The Bromeliad Society Bulletin,* volume 15, p. 23, figs. 1965. *Billbergia marmorata* Lemaire in *L'Illustration Horticole,* volume 2, plate 48. 1855. PLANT propagating by short rhizomes, flowering to 6 dm. high with the inflorescence extended. LEAVES 4-6 in a subcylindric rosette, distichous, 4-5 dm. long, straight, sometimes equalling or exceeding the inflorescence, thick, minutely pale-lepidote beneath and marked with irregular glabrous green or brown bands; sheaths nearly as long as the blades but scarcely wider, suboblong, entire; blades oblong, very broadly rounded or truncate and apiculate, 5-7 cm. wide, laxly serrate with straight spines 1.5 mm. long, green and obscurely punctulate-lepidote above. SCAPE short and erect or longer and decurved, slender, glabrous, red; scape-bracts elliptic, acute, entire, membranaceous, rose, the lower shorter than the internodes, erect and enfolding the scape, the upper slightly longer than the internodes, spreading. INFLORESCENCE very laxly pyramidal, bipinnate at base, simple toward apex, 12-21 cm. long, glabrous; axis straight, rose; primary bracts like the scape-bracts, all large, from slightly shorter than the spikes to exceeding them, spreading or reflexed; spikes very laxly few-flowered, wholly fertile, the lateral ones 3-9 cm. long; rhachis geniculate, rose. FLORAL BRACTS minute, broadly triangular, entire. FLOWERS sessile, to 3 cm. long. SEPALS strongly asymmetric, subelliptic, broadly obtuse and apiculate but unarmed, to 10 mm. long, very short-connate, thick, purple. PETALS erect, ligulate, obtuse to 25 mm. long, purple or blue, bearing two scales at base. STAMENS included. OVARY thick-cylindric or subglobose, scarcely constricted at apex; epigynous tube evident; placentae subcentral; ovules obtuse.

31. Billbergia amoena variety penduliflora

The genus *Billbergia* has two very distinctive types of petals found nowhere else in its subfamily Bromelioideae. The typical subgenus, *Billbergia,* has the petals distinctly irregular in position but only slightly recurved—as is well illustrated in our plate—while the subgenus *Helicodea,* as its name implies, has petals that are regular but coiled back like watch-springs at anthesis. Offhand, *Helicodea* looks like a better genus than *Quesnelia,* but there are some intermediates that spoil the picture. *Billbergia* as a whole is also distinguished by large soft-pointed sepals and pollen grains that lack pores but have a longitudinal fold formed by collapse of a weak area on drying.

Billbergia amoena was described originally in 1818 as a tillandsia and has had a total of seven names applied to it beside the correct one, as well as four varieties separated from it. Making a complete catalogue of its innumerable forms is a task closely akin to bailing a canoe with a teaspoon. Mulford Foster, the author of the variety shown here, writes of his frustration in cultivating one unusual looking rosette after another only to have each bloom prove to be just another *Billbergia amoena.*

One character easily misunderstood in *Billbergia* is the presence or absence of pedicels. Some species have well-developed pedicels while others have one-flowered branches that closely simulate pedicels until one notes the position of the floral bract immediately below the ovary. In our plate there is a good example of the one-flowered branch near the base of the inflorescence.

Billbergia amoena is common in the rainforest of eastern Brazil from Bahia to Santa Catarina. Margaret Mee's plant came from the Serra do Mar in São Paulo while the type of the variety is from Minas Gerais. The two differ in a number of details, such as coloring of bracts and amount of branching but agree in having a pendent inflorescence, so until further fractioning of the species takes place, we will consider them the same variety.

Billbergia Thunberg, *Plantarum Brasiliensium Decas,* volume 3, p. 30. 1821. Stemless usually epiphytic herbs. Leaves forming spreading to slender rosettes; blades usually ligulate. Scape conspicuous, often decurved. Inflorescence simple or compound. Floral bracts conspicuous to minute. Flowers perfect, sessile or pedicellate. Sepals free, erect, subsymmetrical with soft apices. Petals free, large, linear, slightly zygomorphic and suberect or regular and recurved in tight spirals or rarely intermediate, bearing two scales on the inner surface at base. Stamens about as long as the petals but exserted at anthesis, free or the second series short-adnate to the petals; pollen grains with a single longitudinal groove or fold each but no pores. Ovary inferior; epigynous tube evident to very large; ovules numerous, obtuse.

Billbergia amoena (Loddiges) Lindley variety *penduliflora* M. B. Foster in *The Bromeliad Society Bulletin,* volume 12, p. 51. 1962. PLANT propagating by short erect or sometimes long and spreading rhizomes. LEAVES 8-20 in a subbellipsoid rosette, very variable, 3-6 dm. long, from densely lepidote on both sides to nearly glabrous; sheaths large, elliptic, broad; blades ligulate, broadly rounded or acute, apiculate, slightly narrowed toward base, usually all green but sometimes tinged with red or white-spotted, 17-55 mm. wide, serrulate to subentire. SCAPE decurved in the variety *penduliflora,* glabrous, green or red; scape-bracts suberect, elliptic, apiculate, dark red. INFLORESCENCE usually compound but sometimes simple, lax, glabrous, pendent in the variety *penduliflora;* primary bracts like the scape-bracts, large; branches few, few-flowered. FLORAL BRACTS minute, the upper reniform, apiculate. FLOWERS sessile. SEPALS narrowly elliptic, broadly obtuse or acute or minutely apiculate, 20-30 mm. long, normally green with a dark blue apex, white-flocculose at apex. PETALS never spiraled, linear, obtuse, 35-56 mm. long, bearing two fimbriate scales at base, exceeding the stamens but soon recurved, normally green with a dark blue apex. OVARY slenderly cylindric, 10-15 mm. long, sulcate, green.

32. Portea leptantha

The genus *Portea* is distinguished by its slender pedicels, mucronate sepals and multiporate pollen and is restricted to eastern Brazil from Paraiba to Rio and Minas. Some billbergias have slender pedicels but their sepals are never mucronate. Some aechmeas have both slender pedicels and mucronate sepals but their pollen is not more than 4-porate and they are confined to western America. For quick recognition their flowers are much smaller than any in *Portea*.

Of the six species, *Portea leptantha* is the northernmost and also the largest and most striking. It is not very common in cultivation but there are some magnificent specimens in the botanic gardens in München (Munich), Germany. In order to do full justice to the inflorescence, the leaves are omitted from this plate but they are spine-tipped and coarsely serrate as might be expected in the "Nordeste." The leaves of *Hohenbergia horrida* are similar but considerably more ferocious.

Margaret Mee's exuberant specimen comes from Gravatá (a popular name for bromeliad) near Russinha in eastern Pernambuco and not far from Tapera, the type locality.

Portea K. Koch, *Index Seminum in Horto Botanico Berolinensi, Anno 1856 Collectorum,* Appendix, p. 7. 1857. Perennial herbs. Leaves rosulate, lepidote; blades ligulate, serrulate. Scape central; scape-bracts brightly colored. Inflorescence compound. Flowers slenderly pedicellate perfect. Sepals more or less connate, strongly asymmetric, long-mucronate. Petals free, symmetric, exceeding the stamens and pistil, bearing two fimbriate scales at base. Stamens of the second series highly adnate to the petals; pollen with more than five pores. Ovary wholly inferior; placentae various; ovules numerous, long-caudate. Fruit fleshy.

Portea leptantha Harms in *Notizblatt,* volume 10, p. 786. 1929. PLANT flowering 16 dm. or higher. LEAVES to 9 dm. long, rigid; sheaths broadly elliptic, 2 dm. long, covered with minute appressed brown scales, sometimes tinged with purple toward apex; blades broadly acute and then attenuate into a long subulate apex, to 10 cm. wide at base, green, nearly glabrous above, minutely cinereous-lepidote beneath, laxly serrate with curved brown spines 4 mm. long. SCAPE erect, about 1 cm. in diameter at base, sparsely flocculose or glabrous; scape-bracts subfoliaceous, erect, the upper exceeding the internodes. INFLORESCENCE much-branched, broadly paniculate, 25-40 cm. in diameter, subglabrous or glabrous; primary bracts narrowly lanceolate, much shorter than the corymbose branches, the other bracts gradually smaller; ultimate branches few-flowered, geniculate, very slender. FLORAL BRACTS ovate, long-mucronate, to 8 mm. long; pedicels to 12 mm. long, slender. SEPALS narrowly obovate, slenderly mucronulate, 15 mm. long, very short-connate. PETALS linear, 20-35 mm. long, bearing two scales at base. OVARY obconic, 7-8 mm. long.

Plates

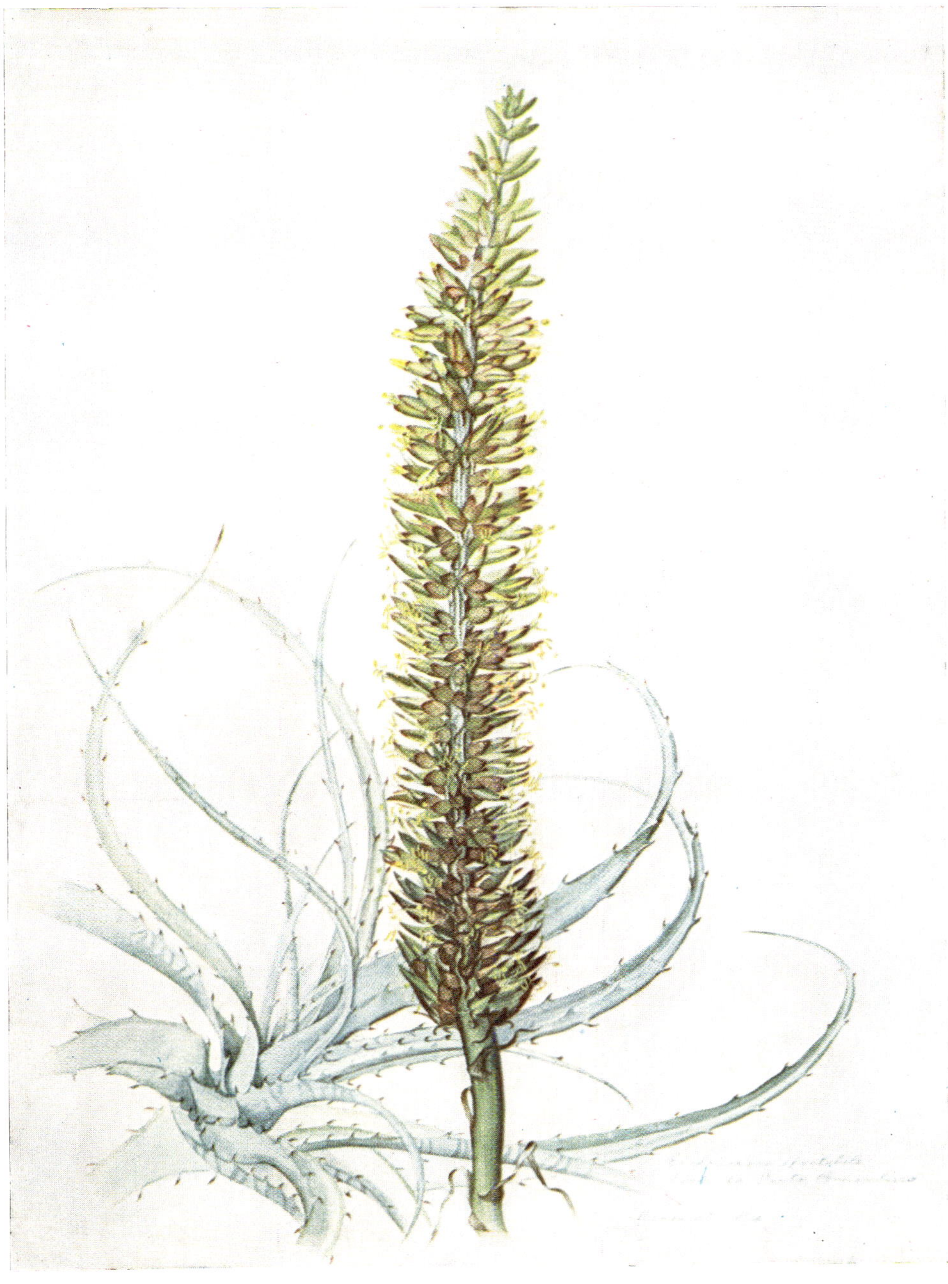

1. *Encholirium spectabile*

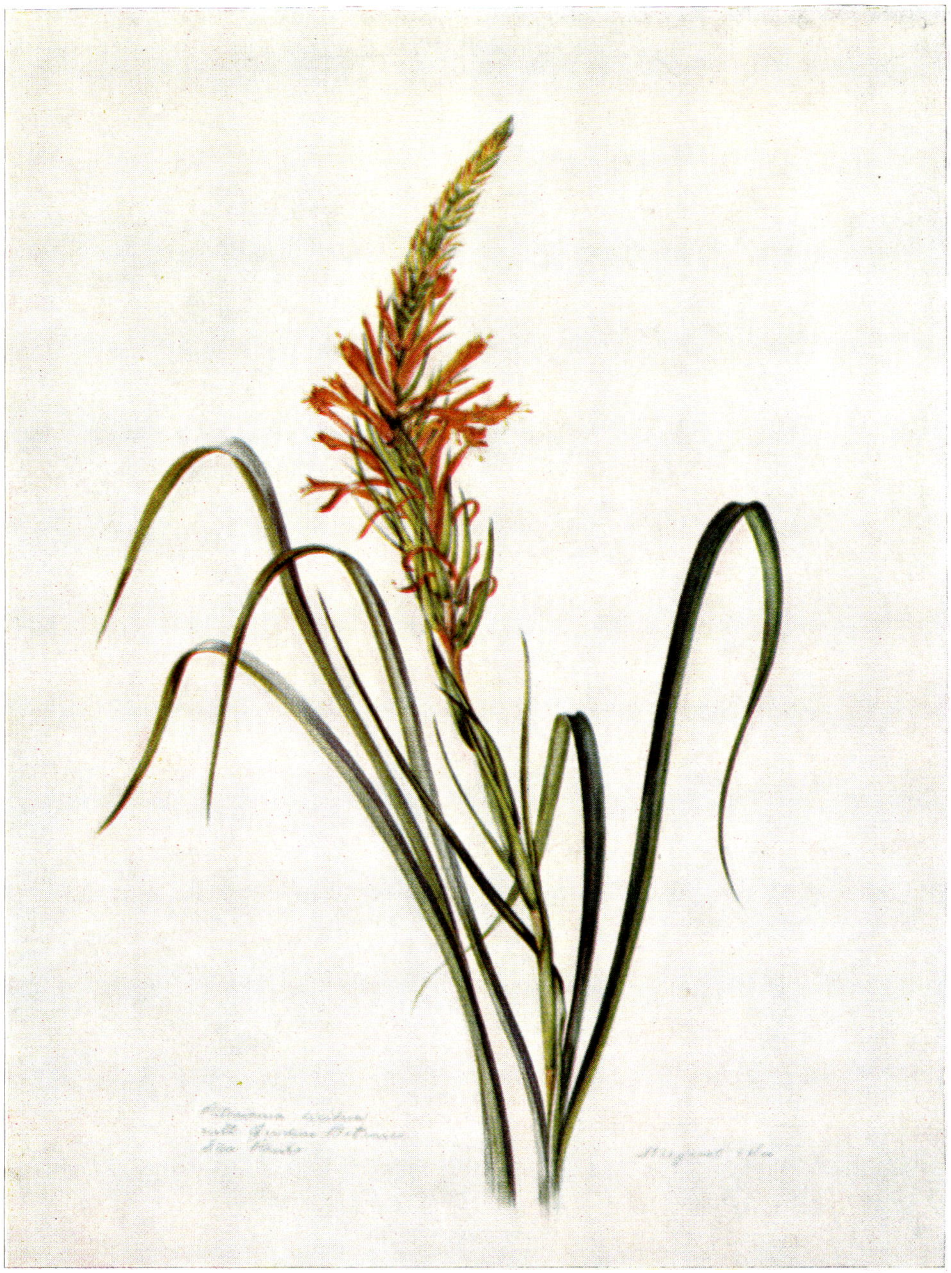

2. *Pitcairnia decidua*

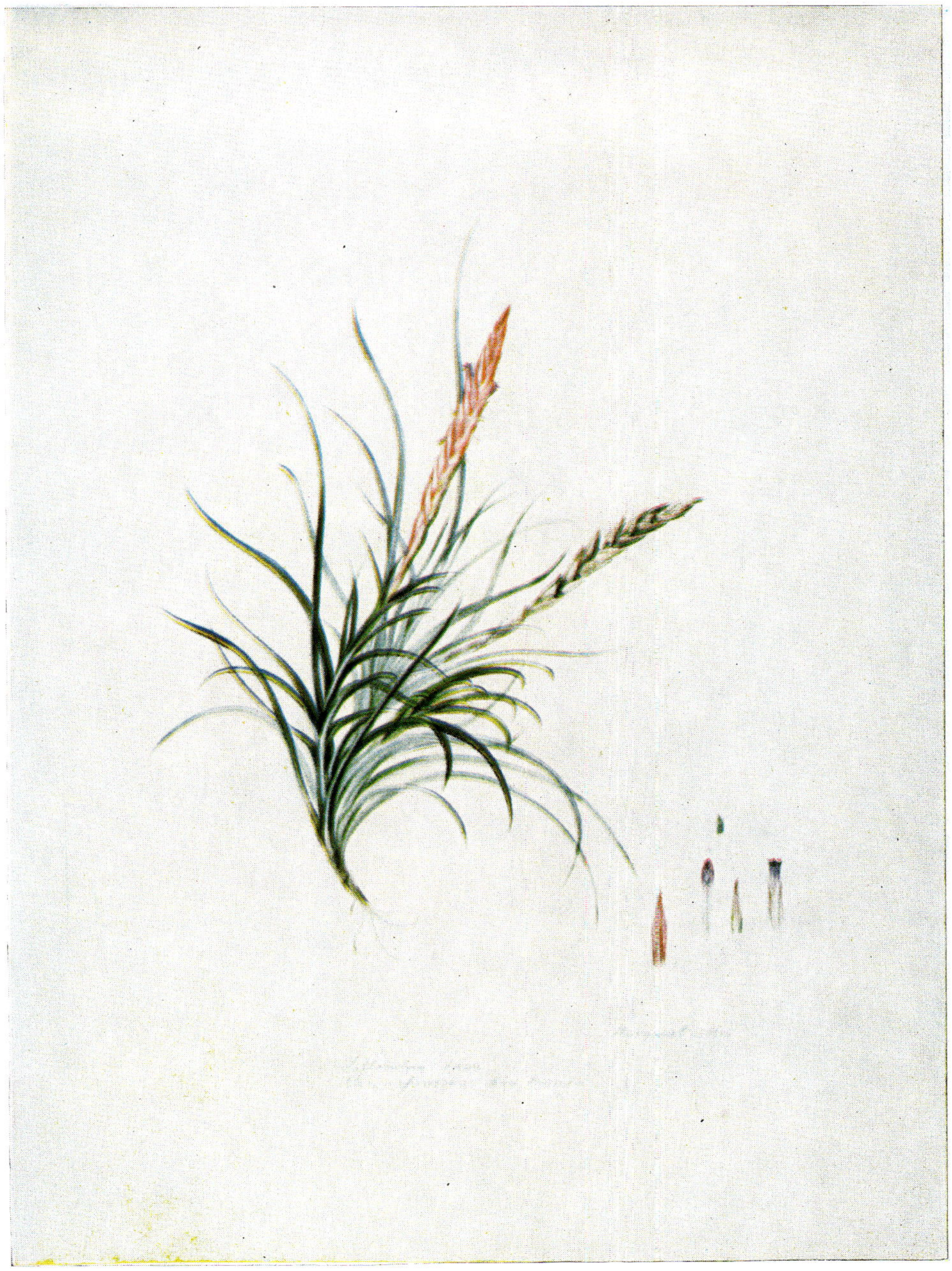

3. *Tillandsia dura*

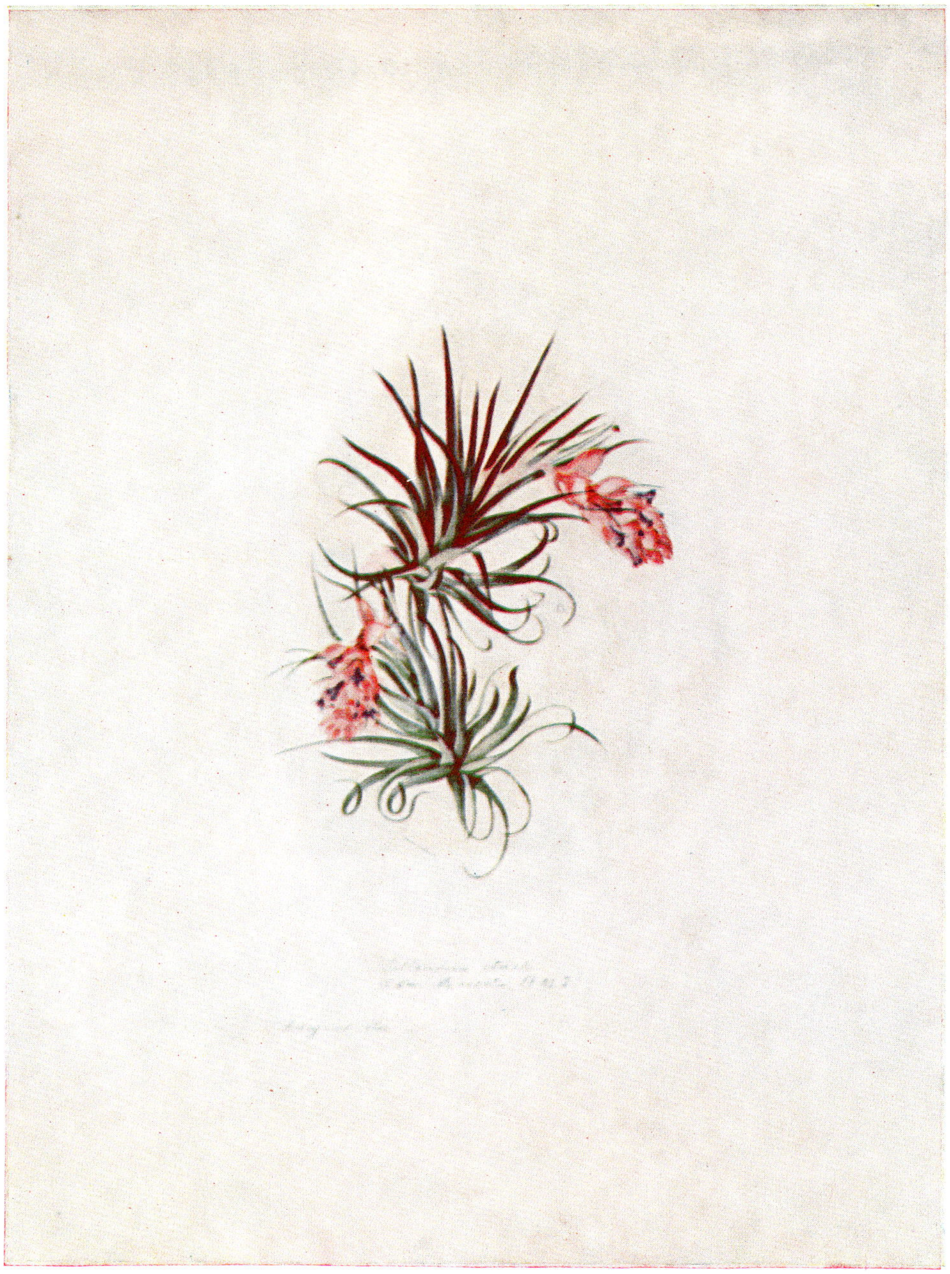

4. Tillandsia stricta

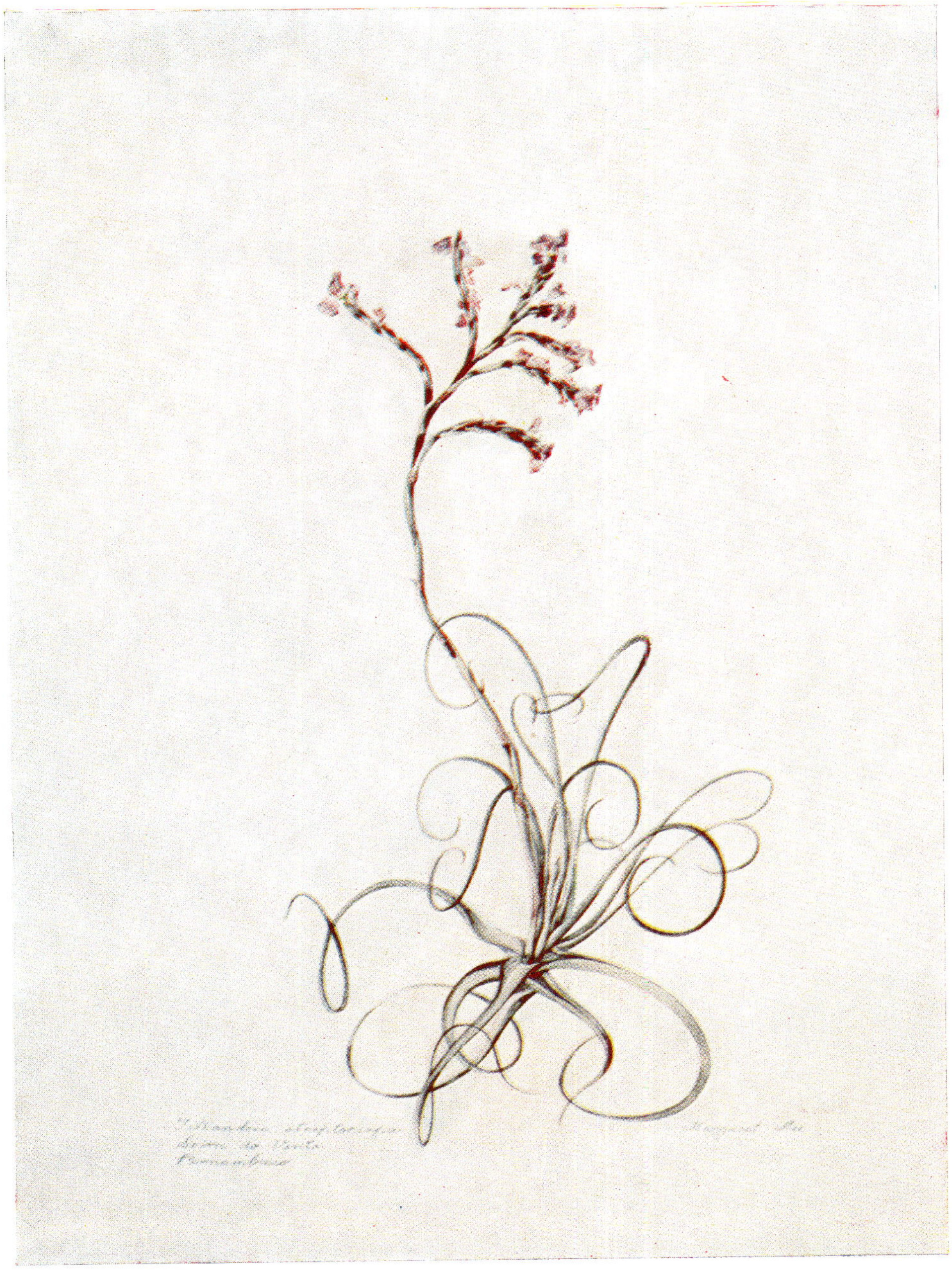

5. *Tillandsia streptocarpa*

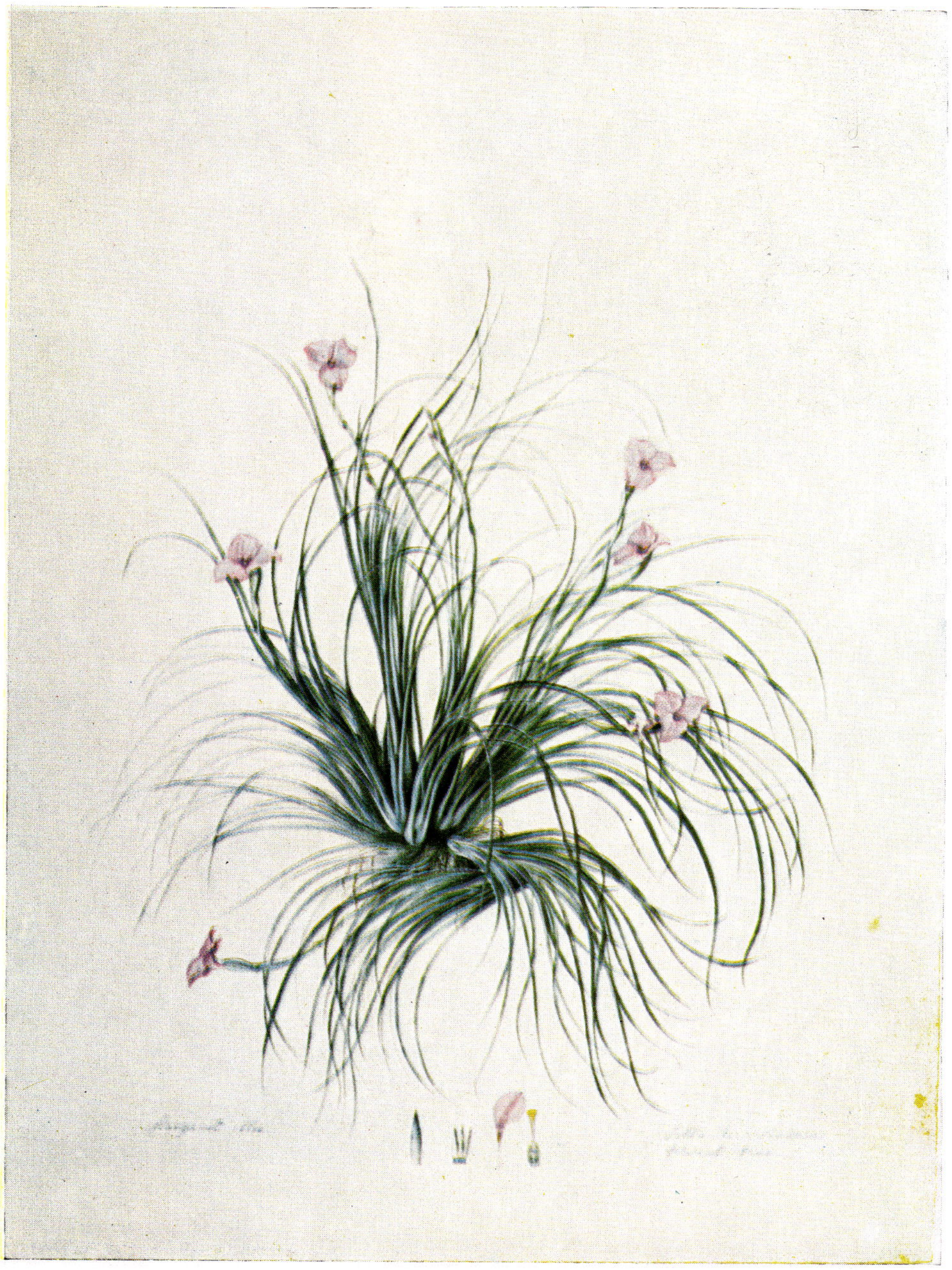

6. *Tillandsia linearis*

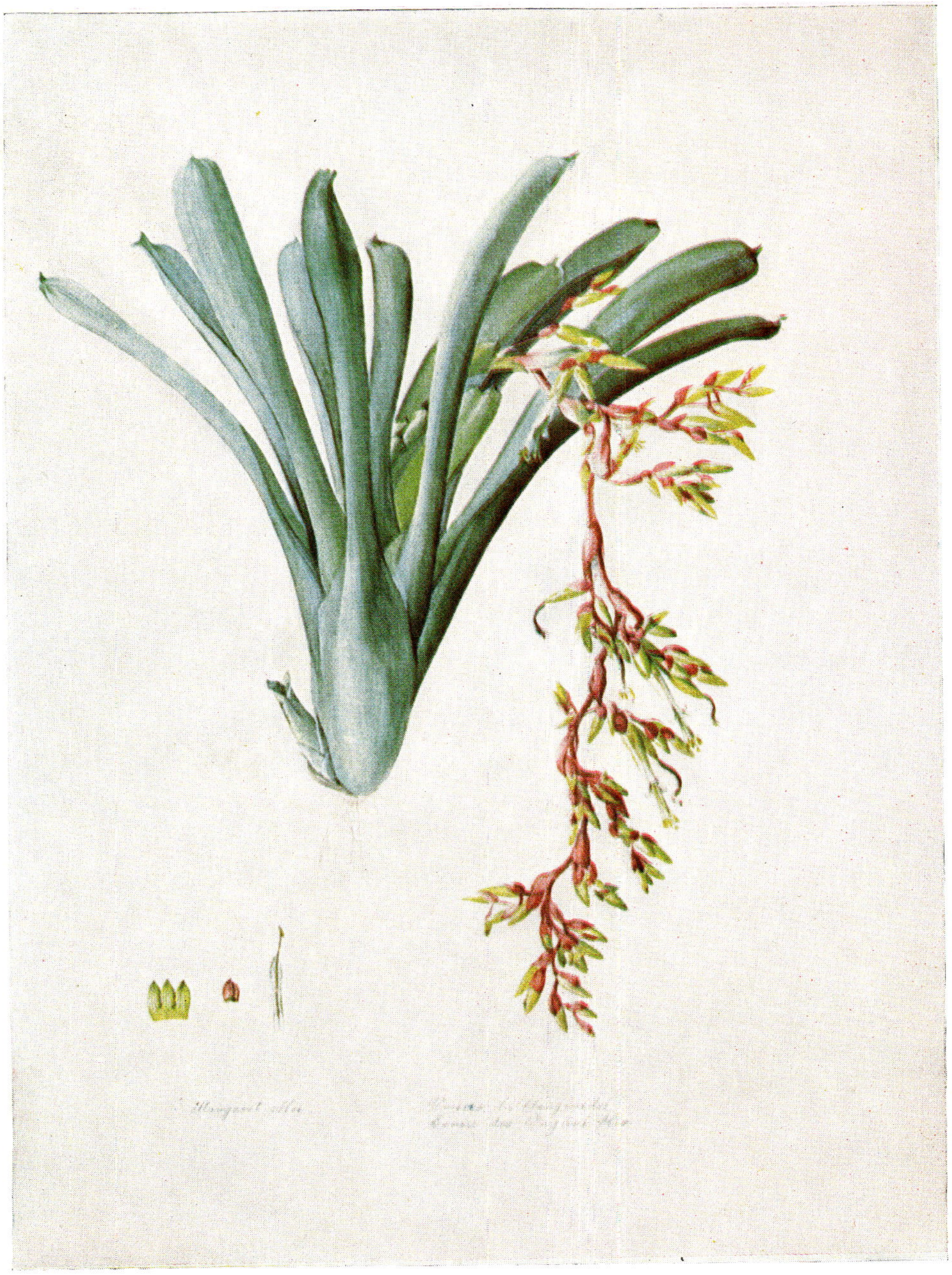

7. *Vriesea billbergioides*

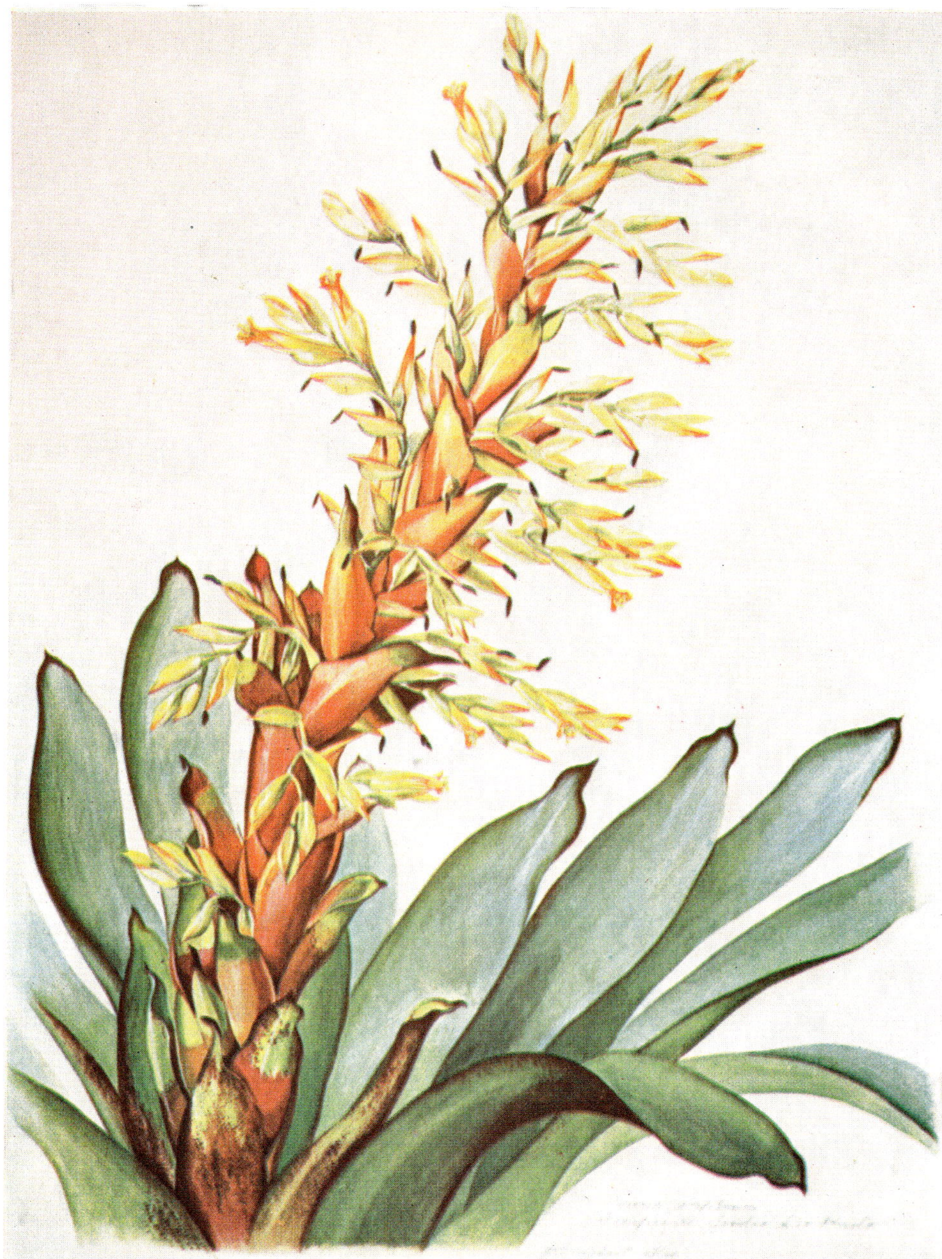

8. Vriesea sceptrum

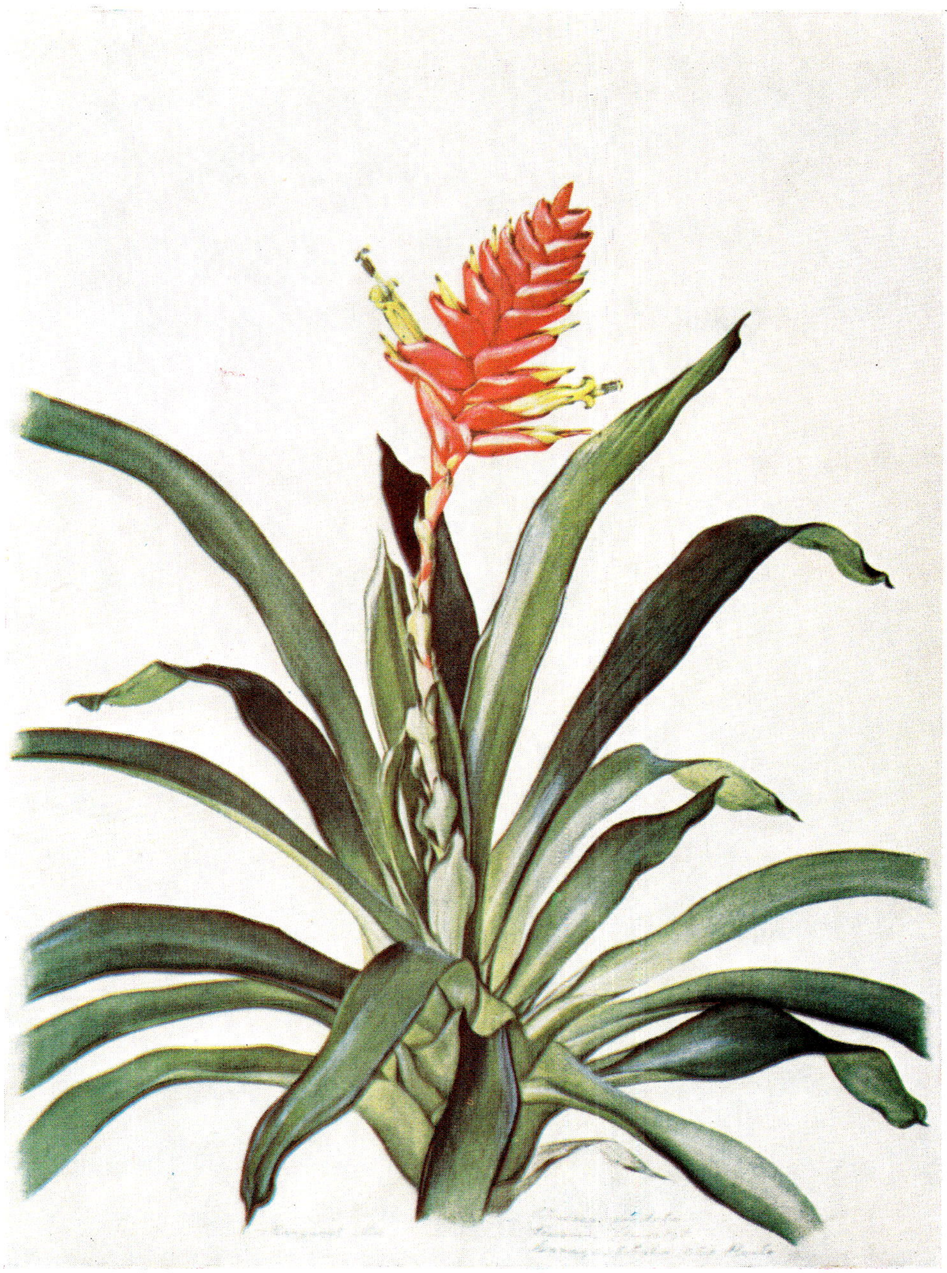

9. *Vriesea gradata*

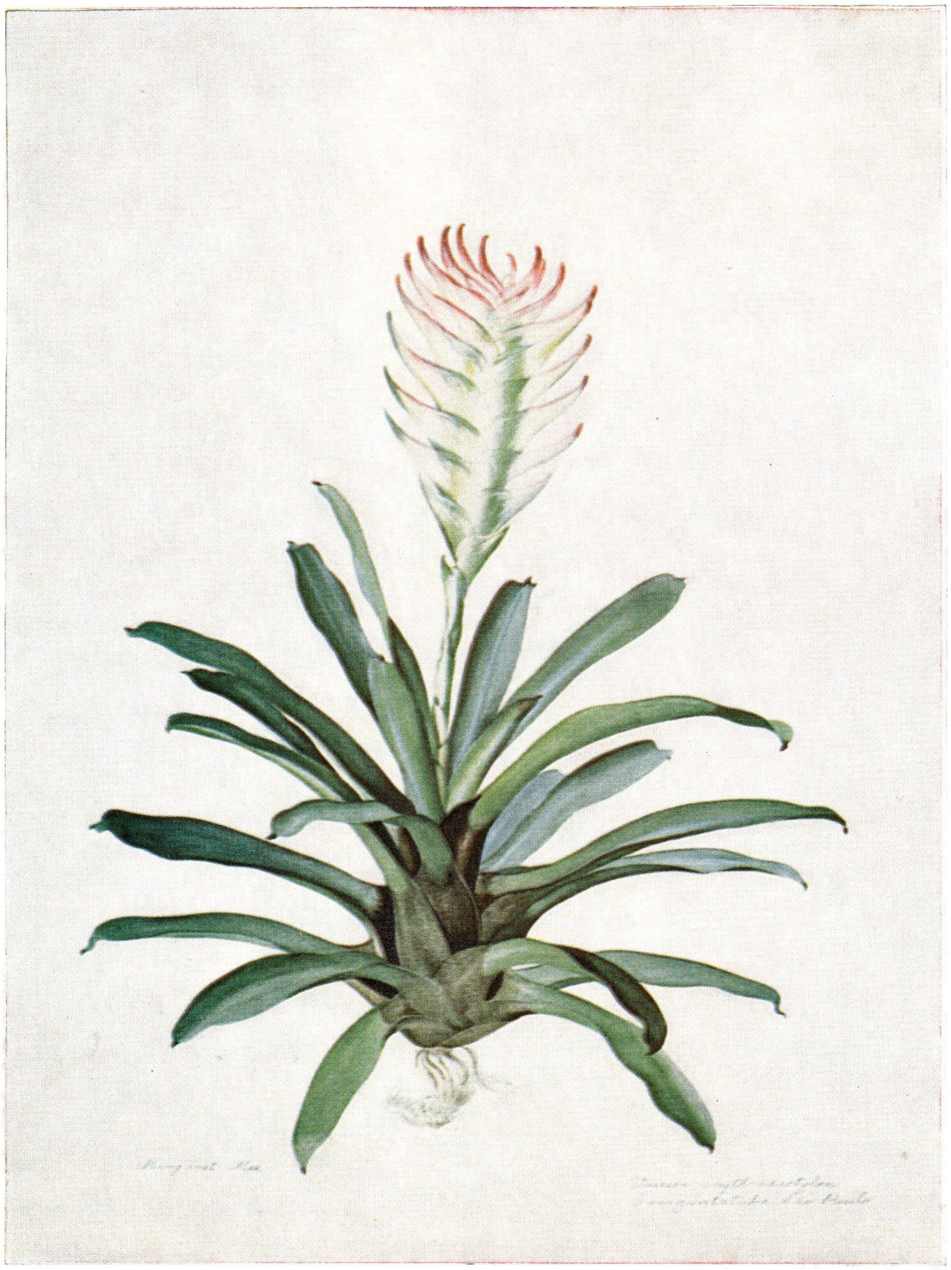

10. Vriesea erythrodactylon

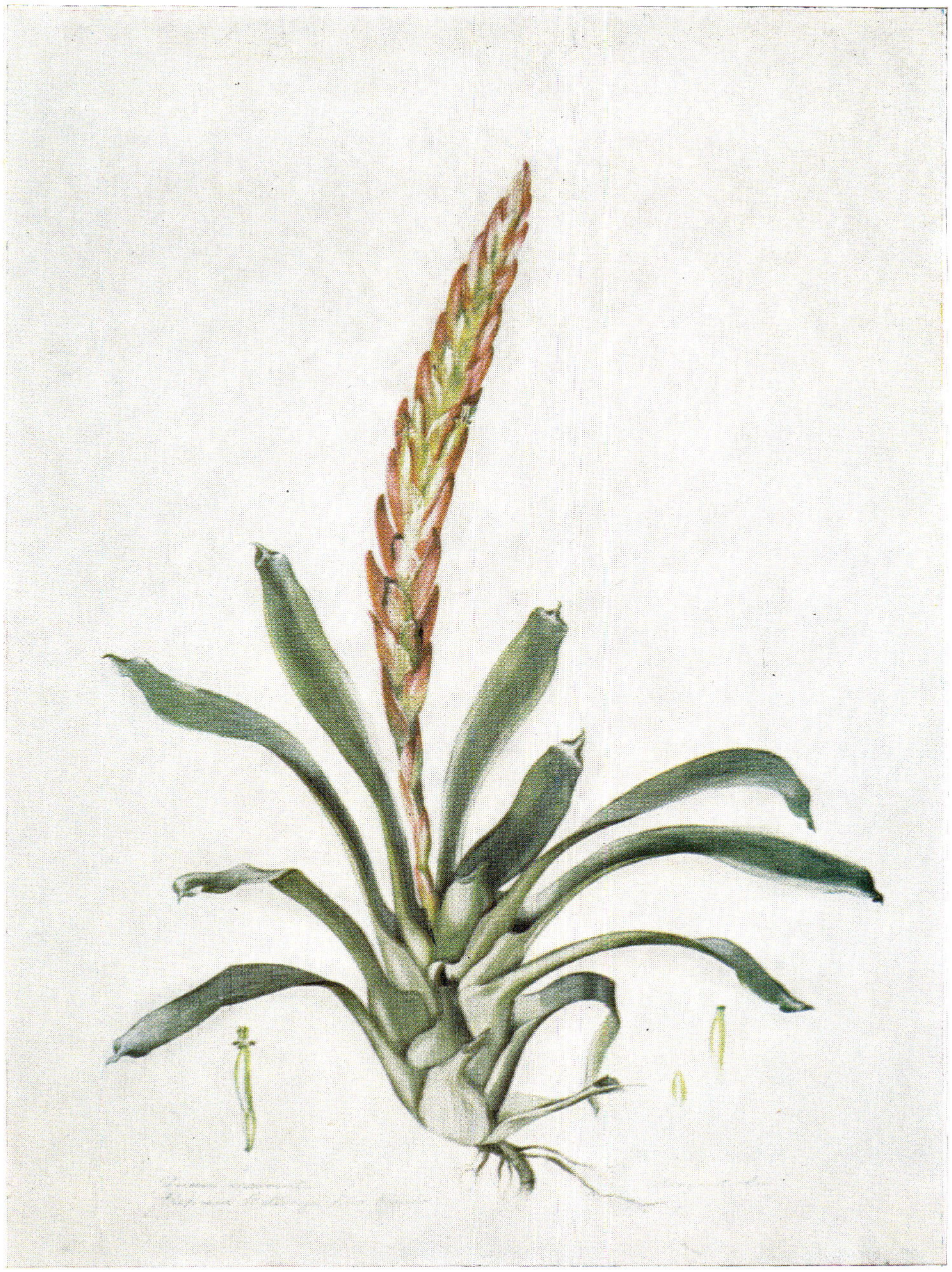

11. *Vriesea incurvata*

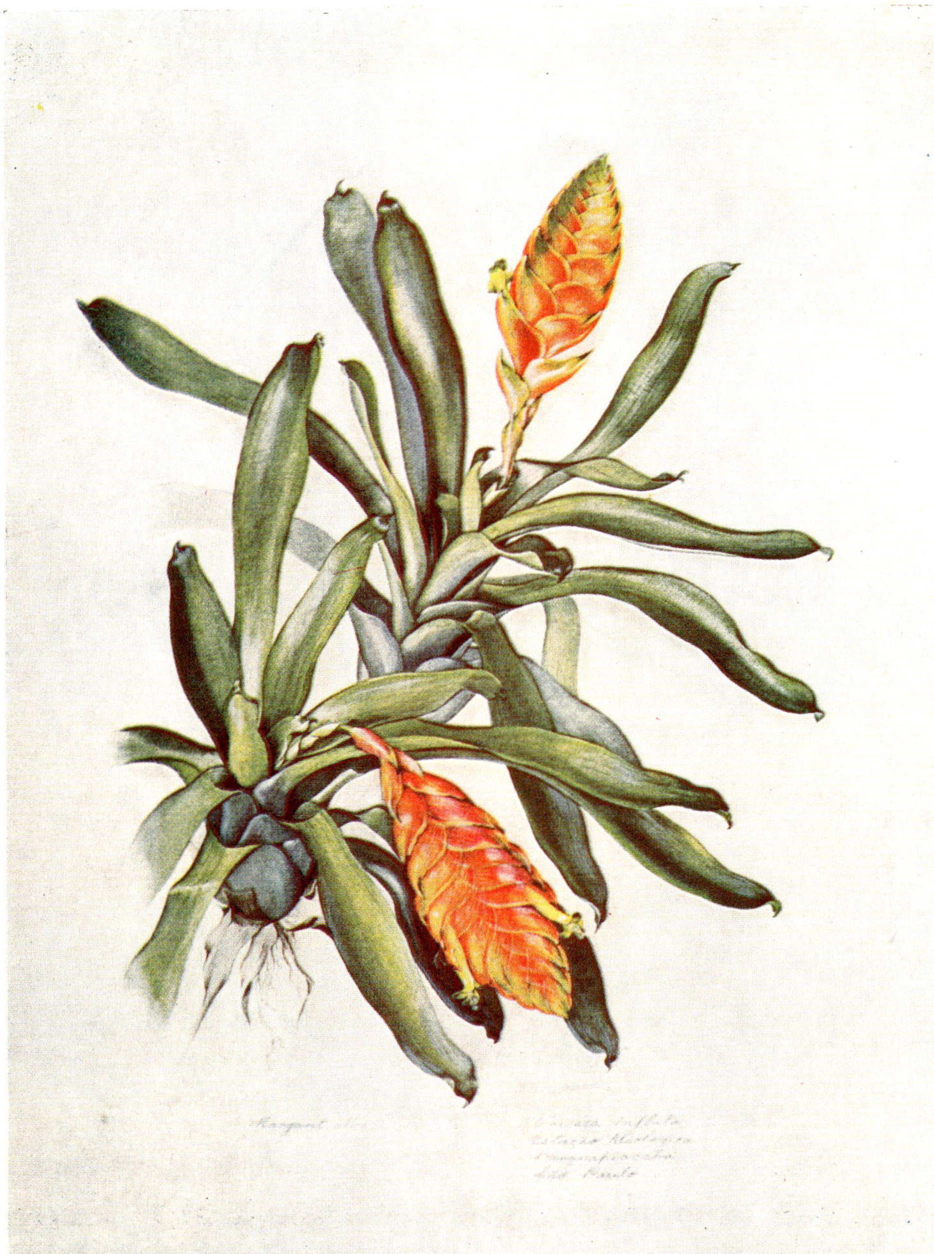

12. *Vriesea inflata*

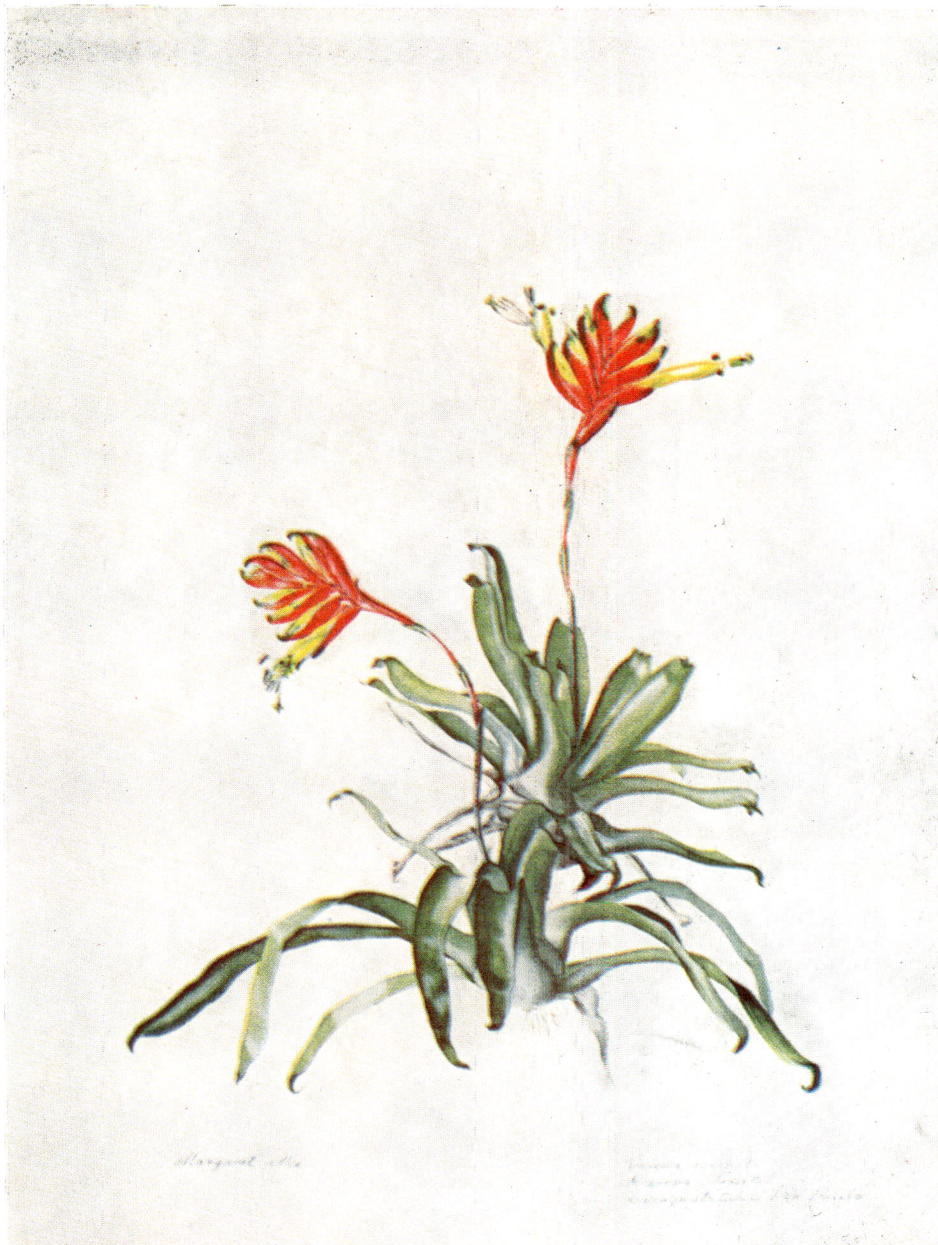

13. *Vriesea carinata*

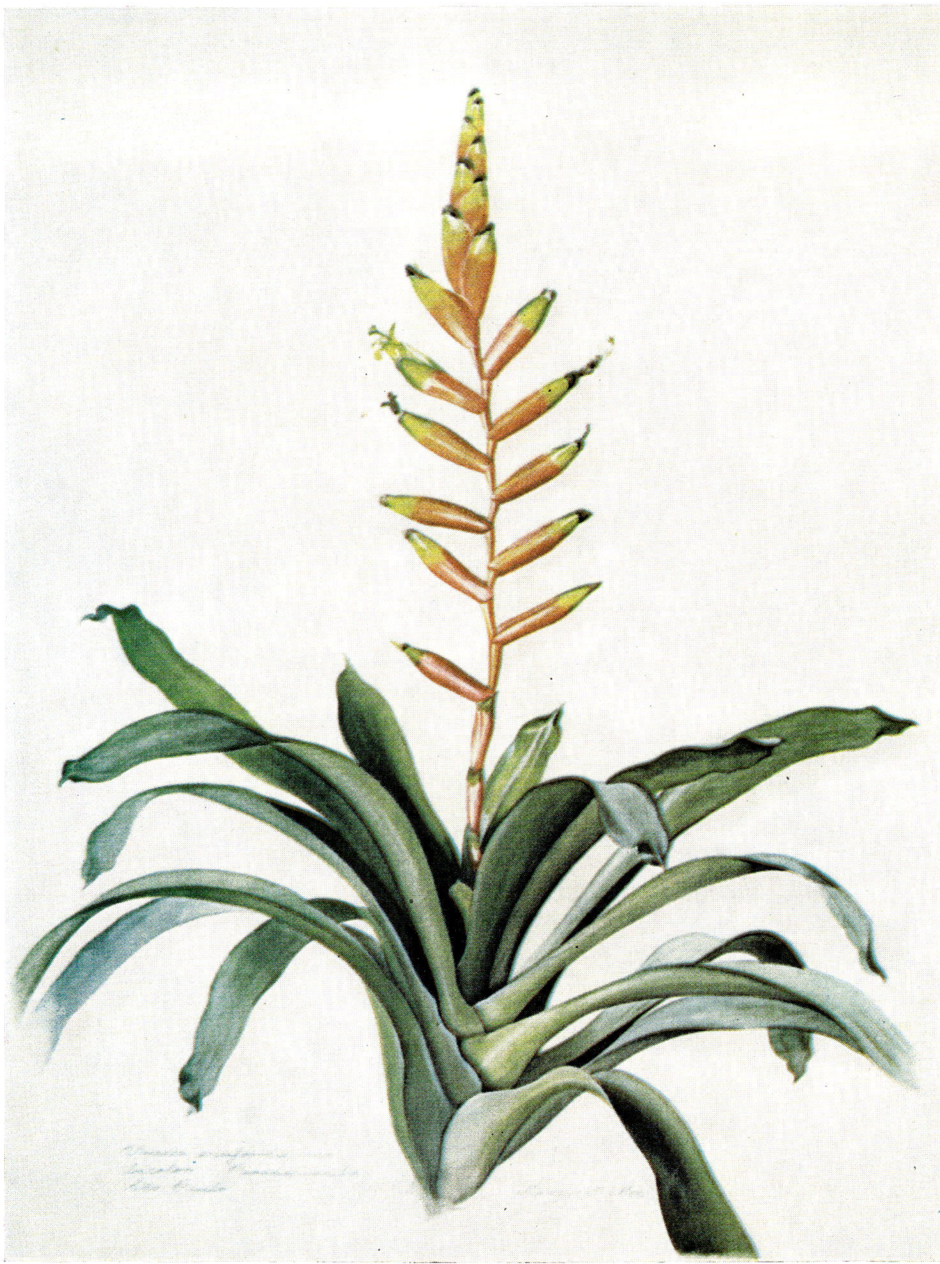

14. *Vriesea ensiformis* var. *bicolor*

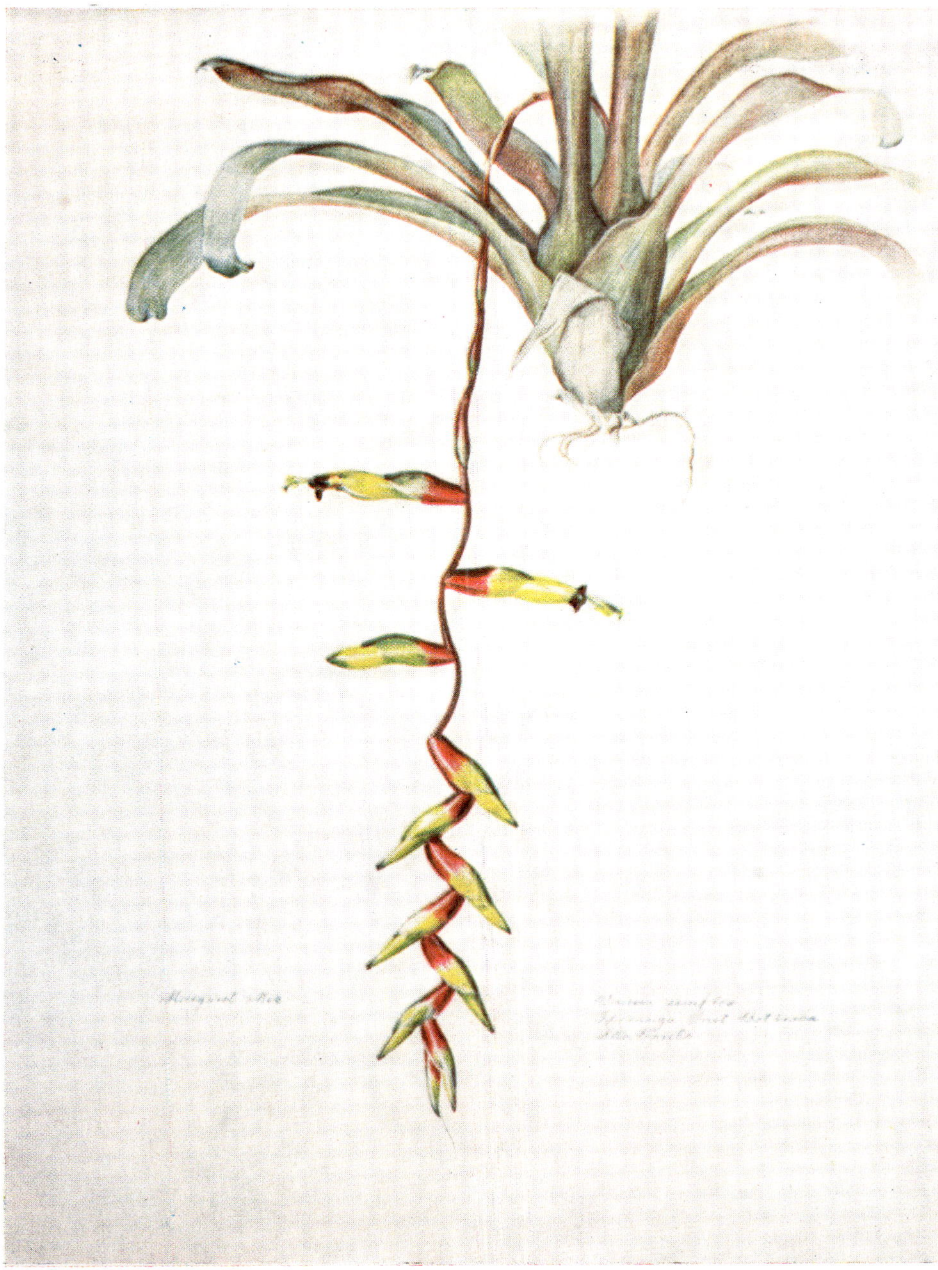

16. Vriesea simplex

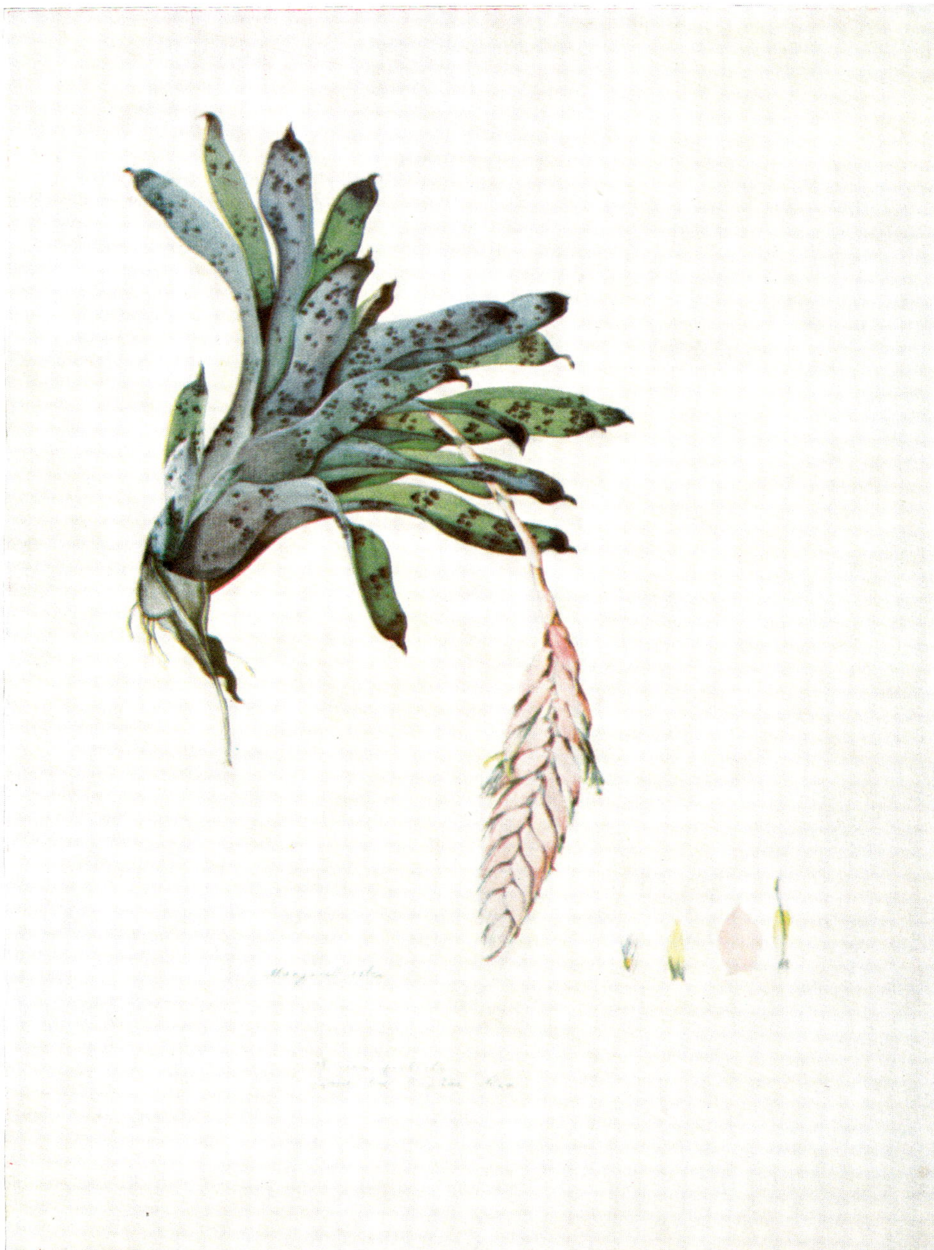

15. *Vriesea guttata*

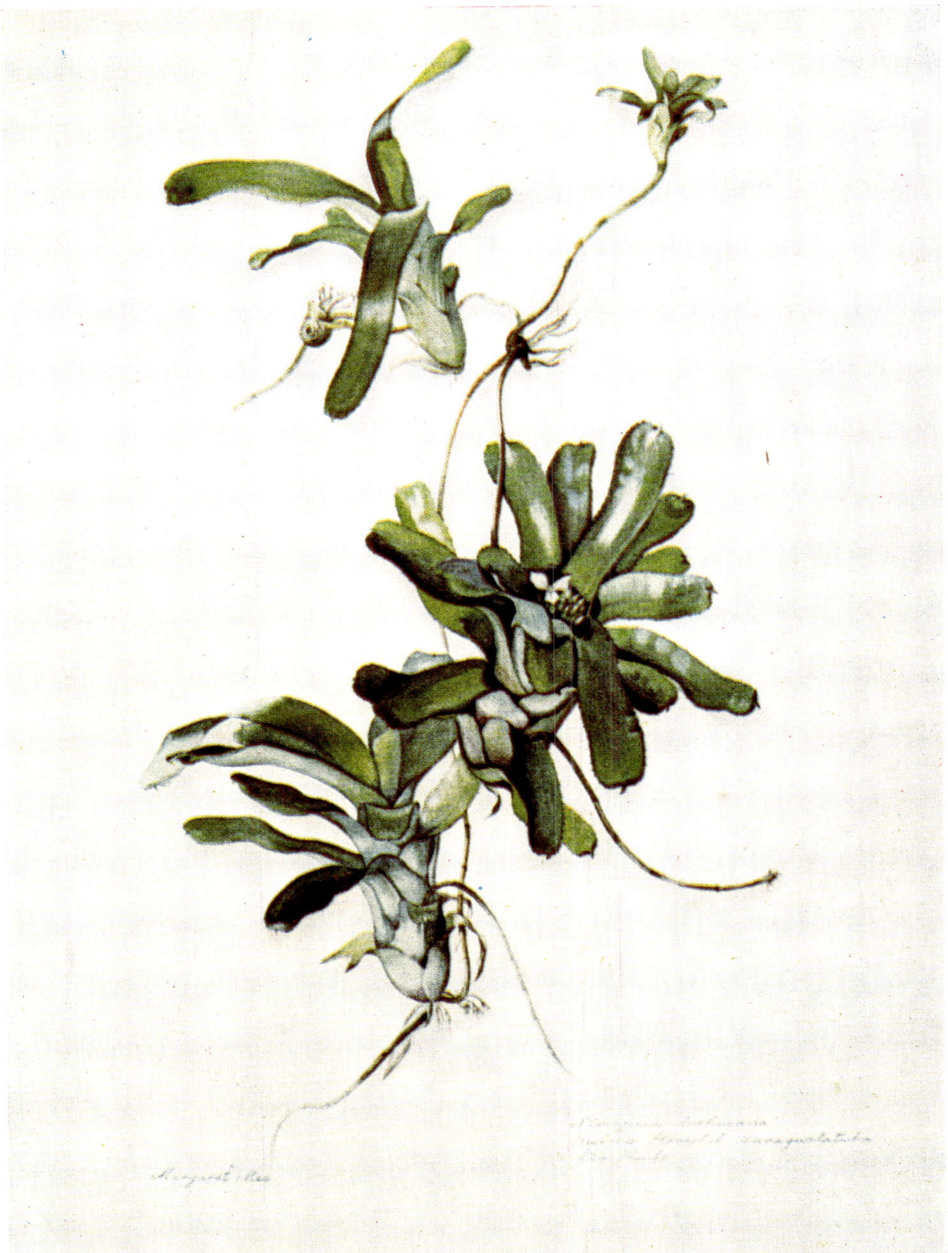

17. Neoregelia hoehneana

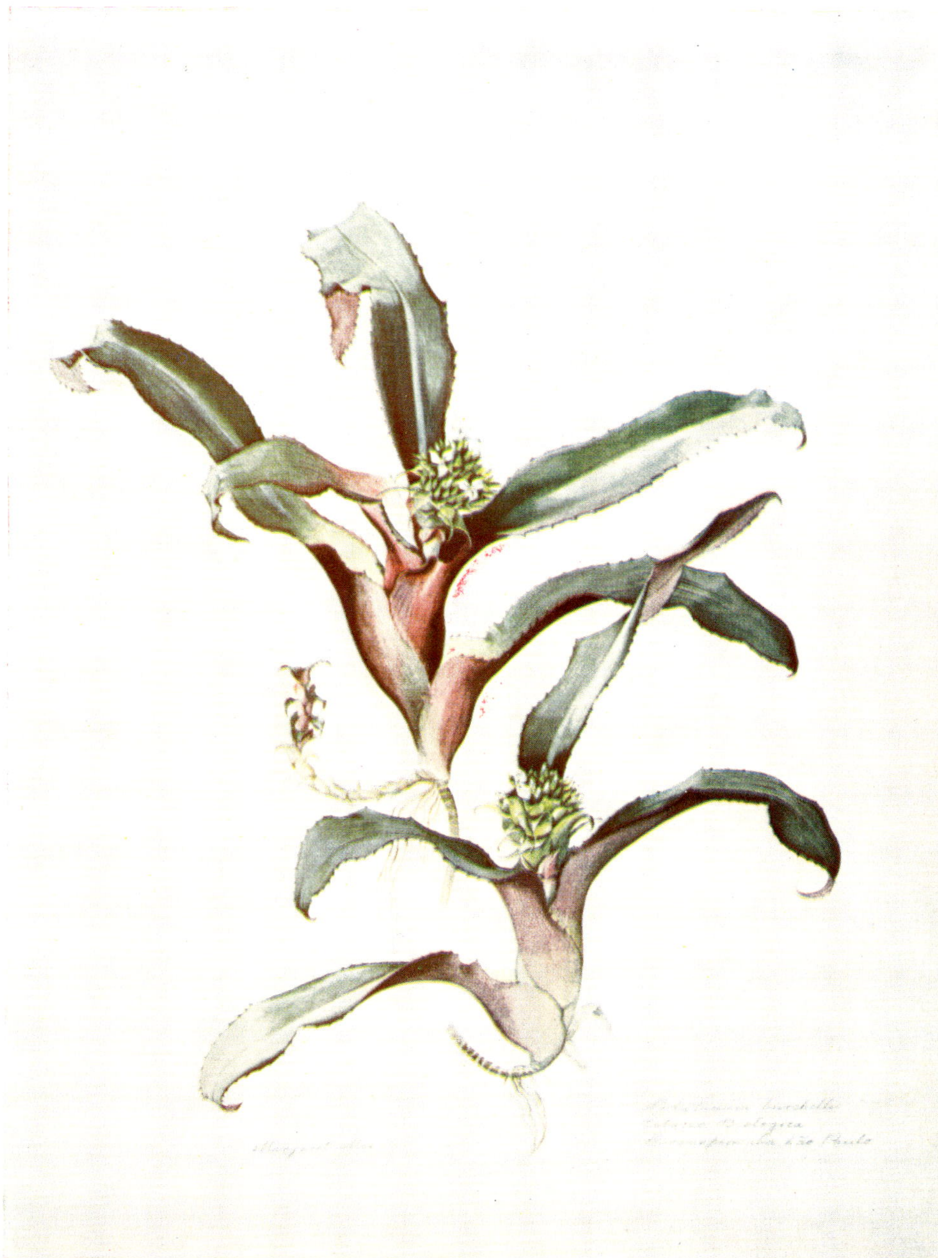

18. *Nidularium burchellii*

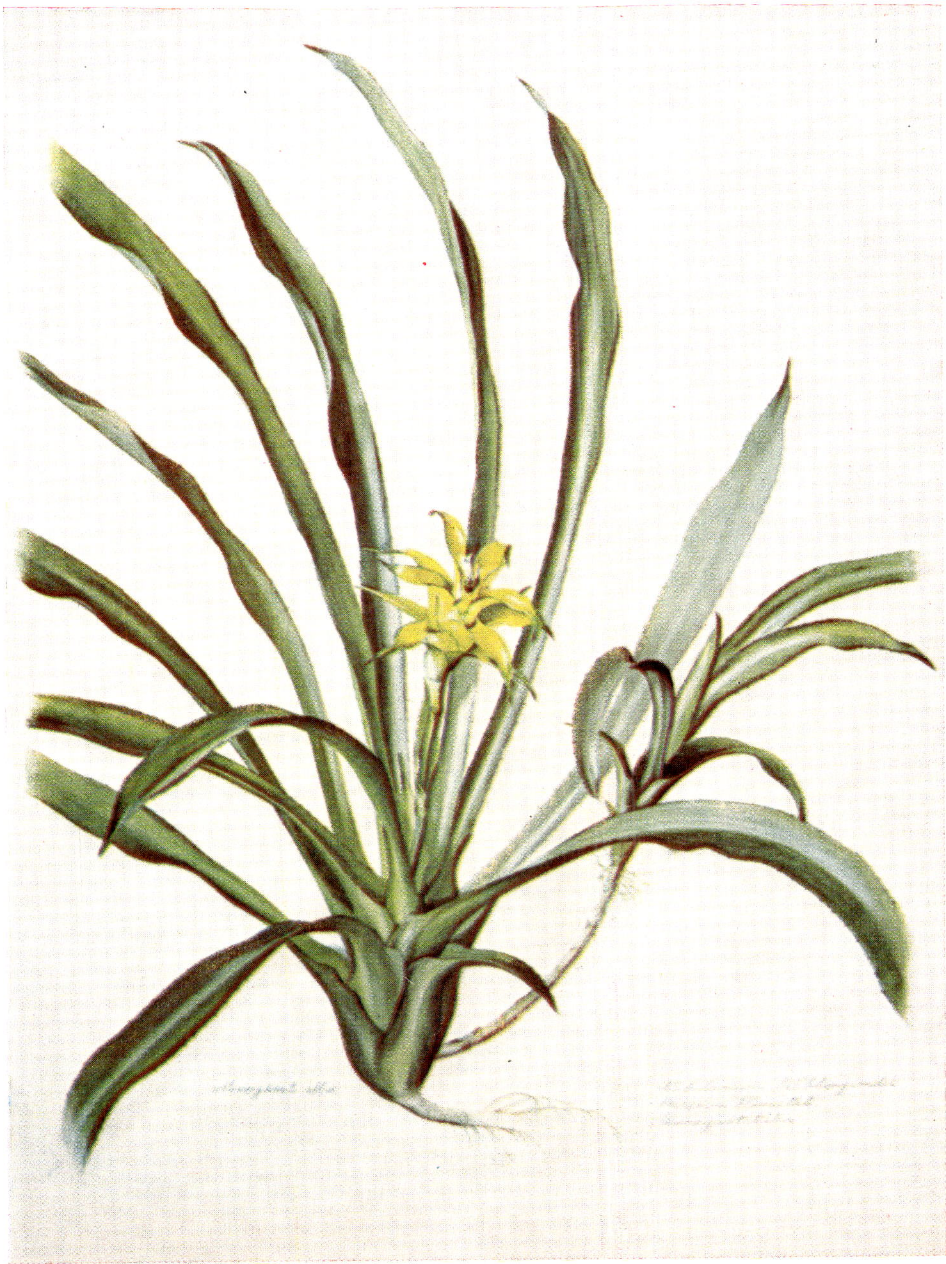

19. Nidularium billbergioides

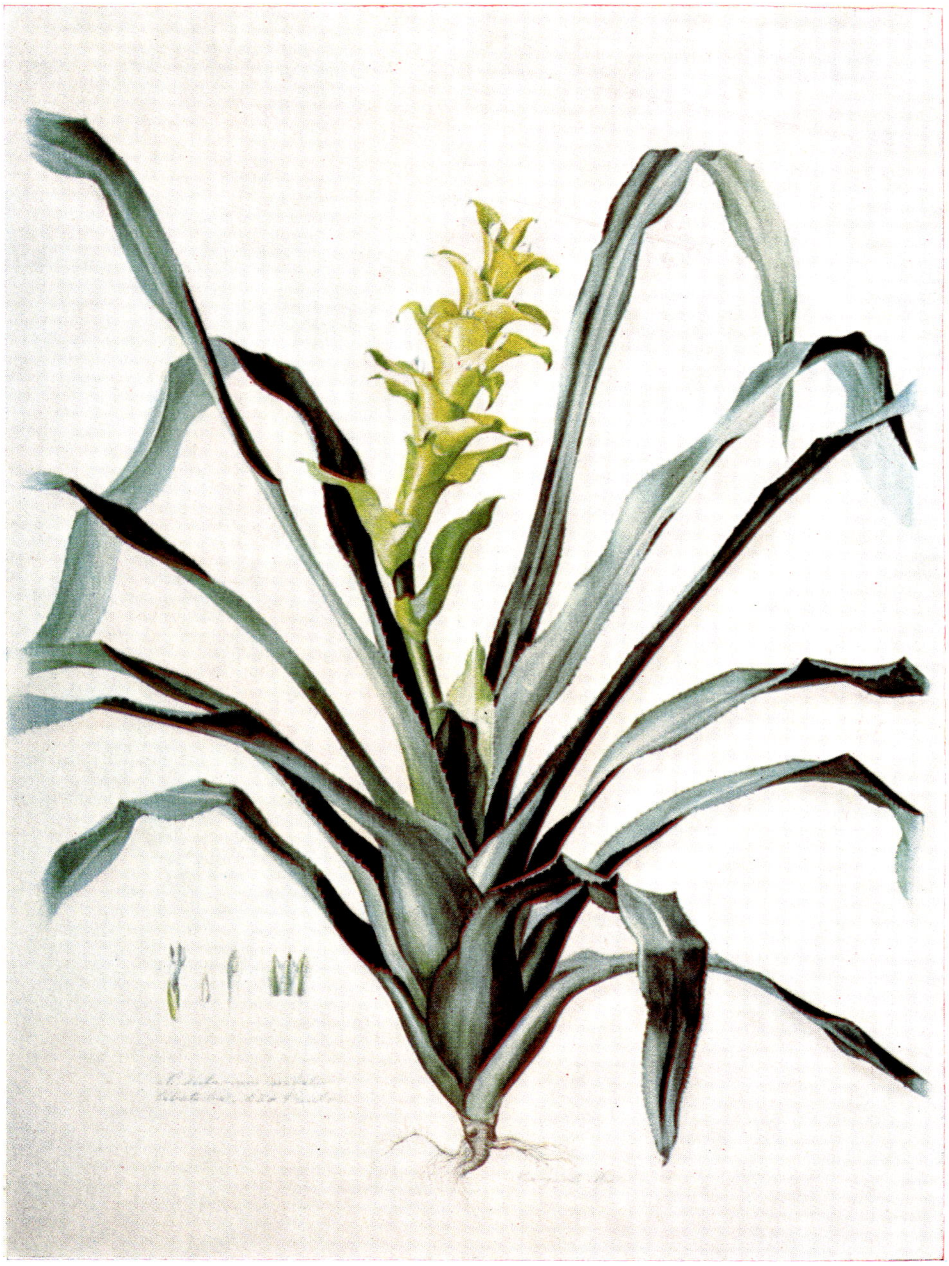

20. *Nidularium seidelii*

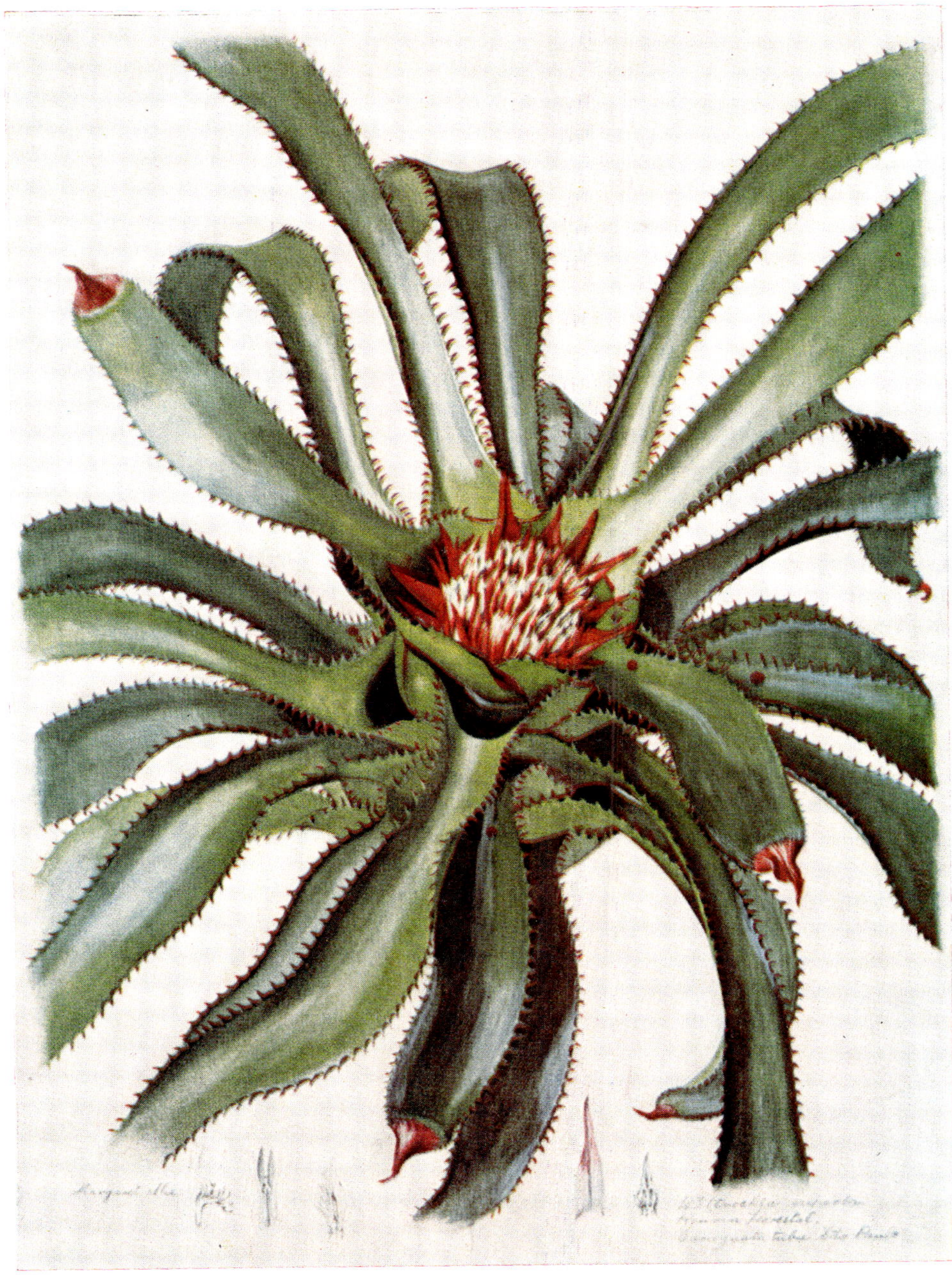

21. *Wittrockia superba*

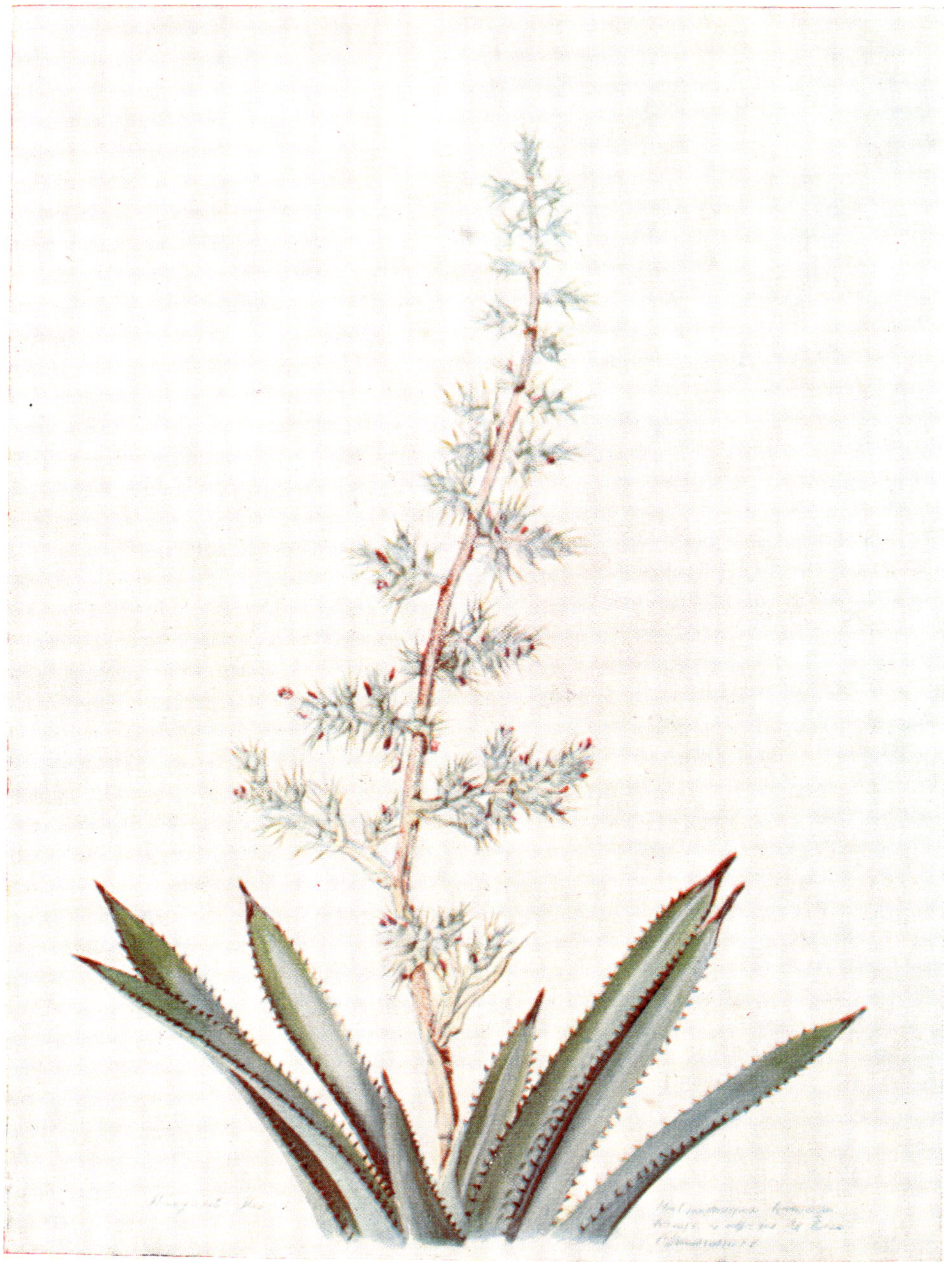

22. *Hohenbergia horrida*

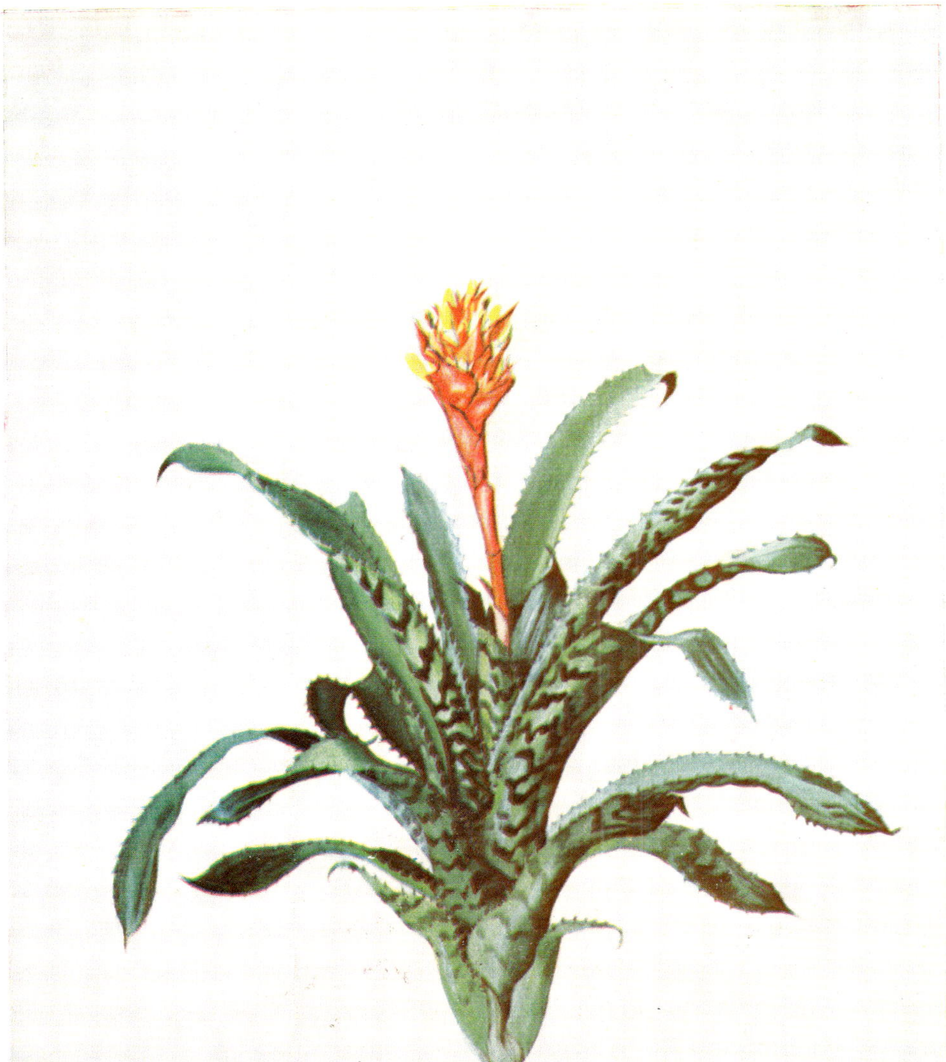

23. *Aechmea orlandiana*

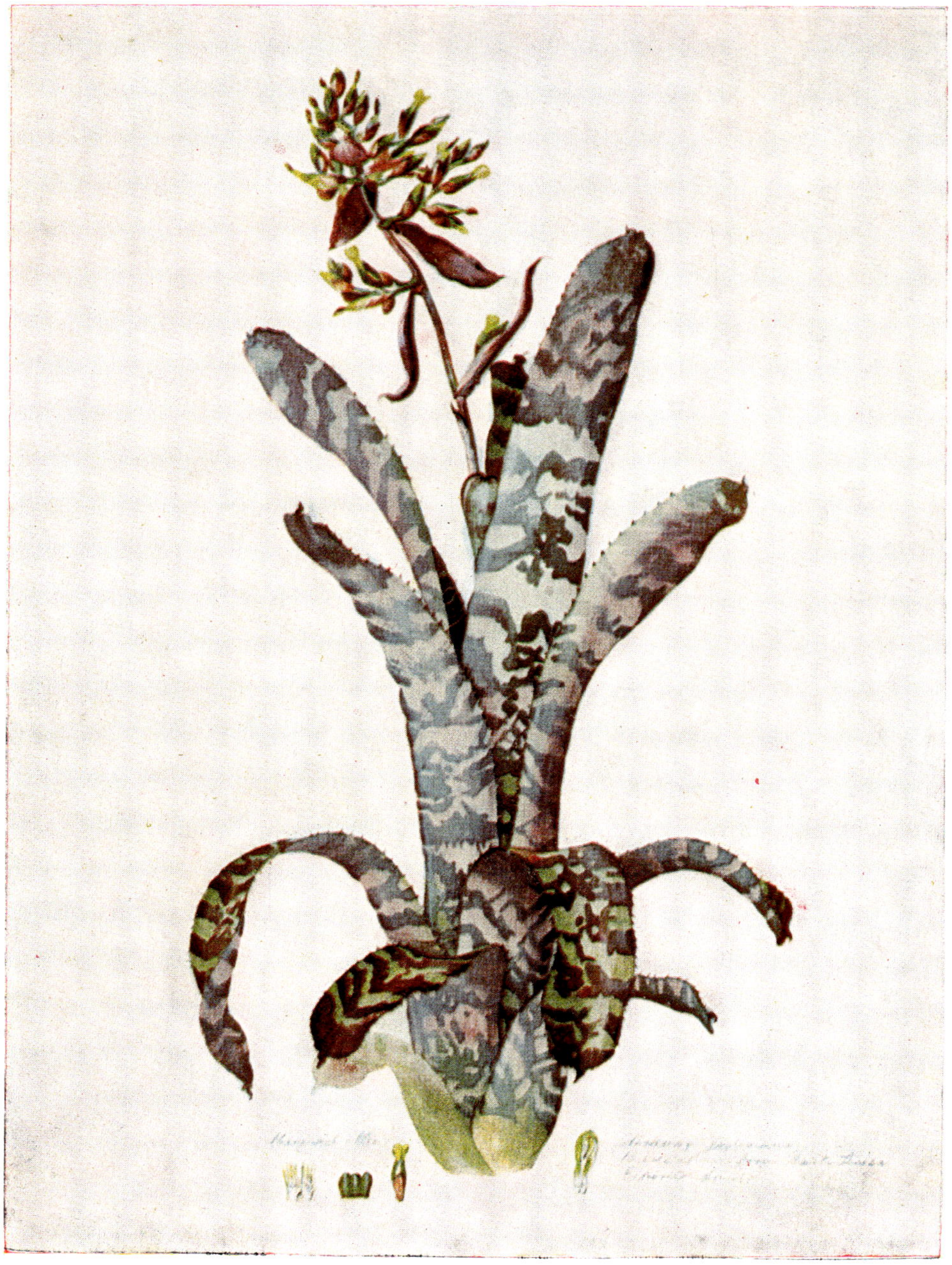

24. *Aechmea fosteriana*

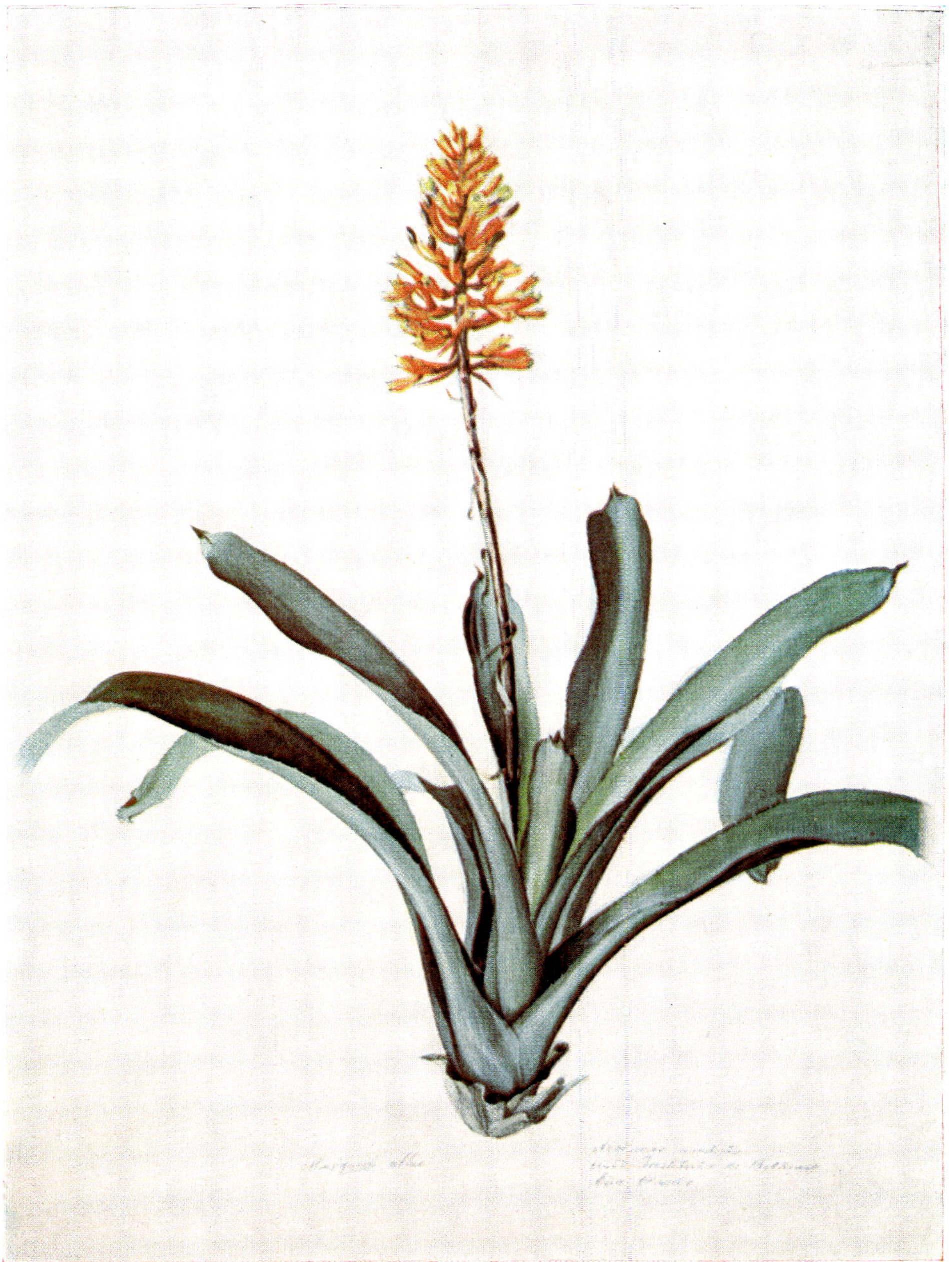

25. *Aechmea caudata*

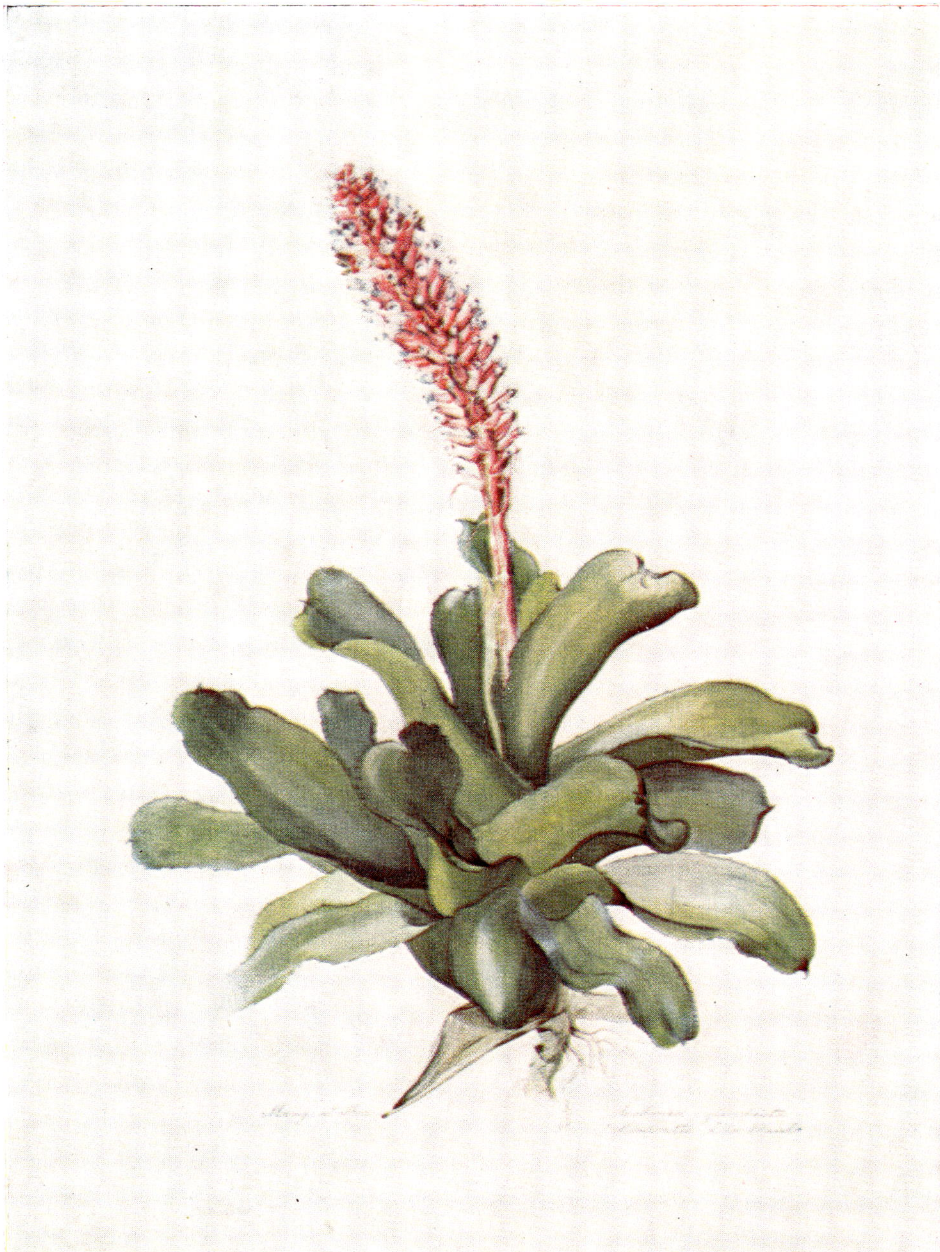

26. Aechmea cylindrata

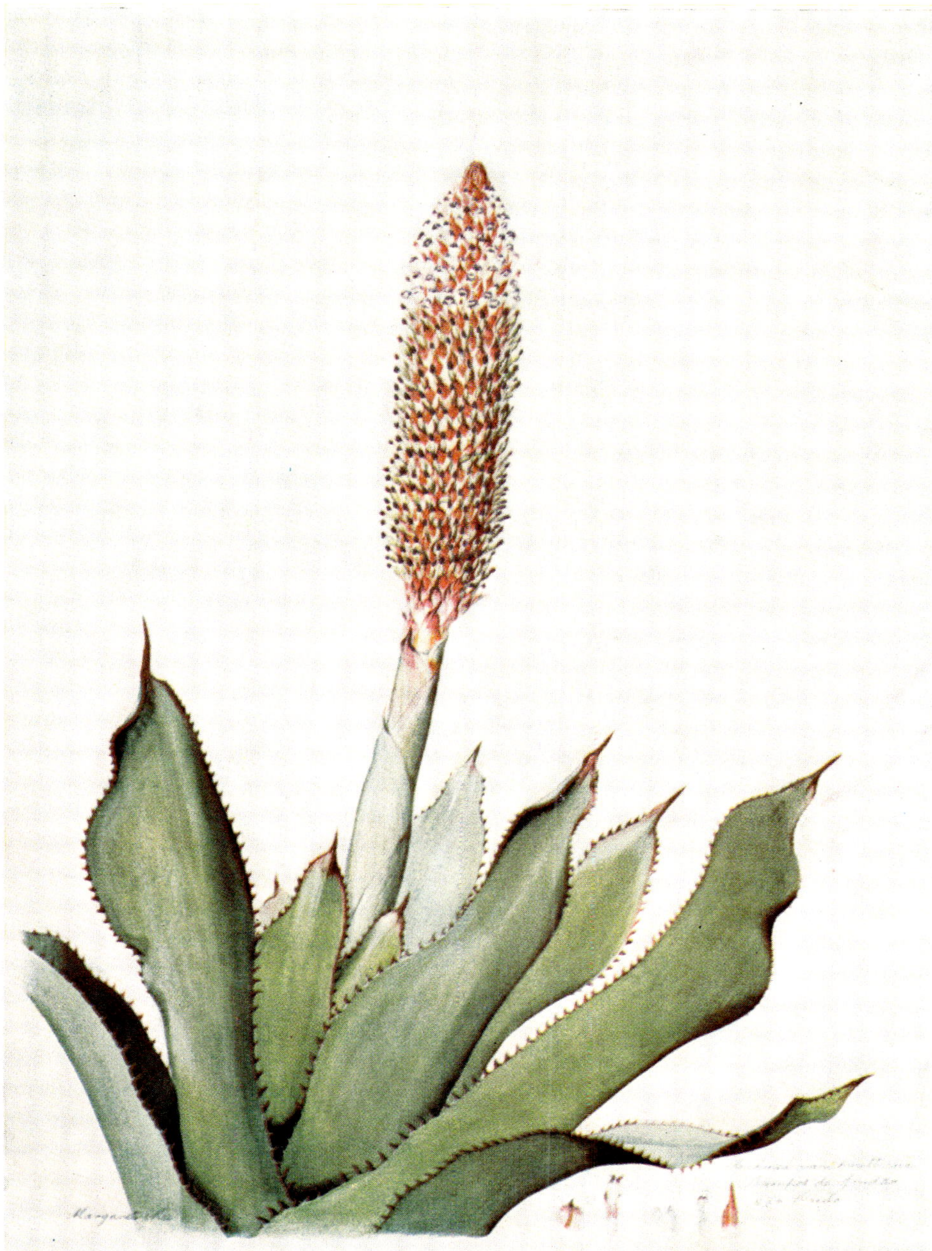

27. *Aechmea vanhoutteana*

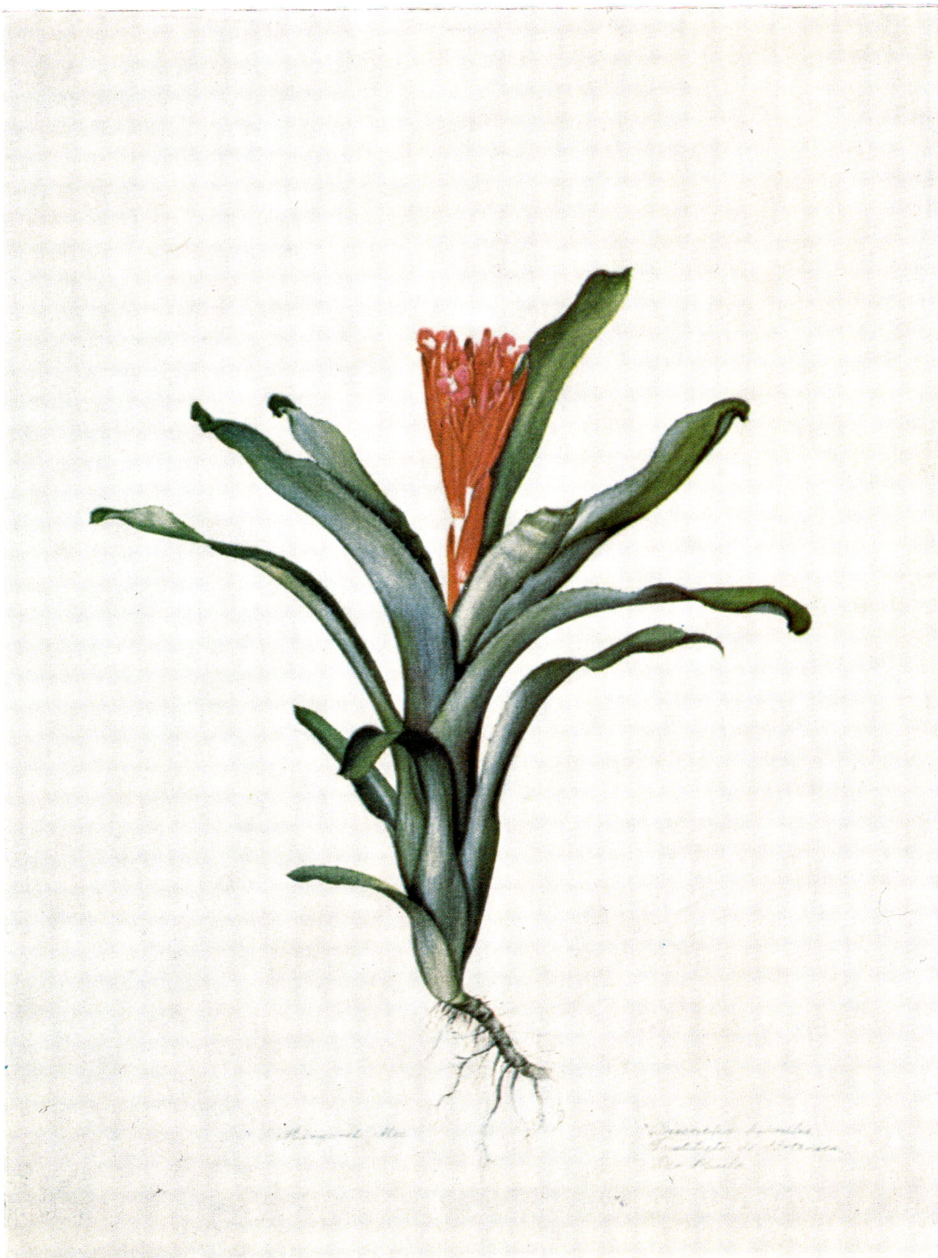

28. *Quesnelia humilis*

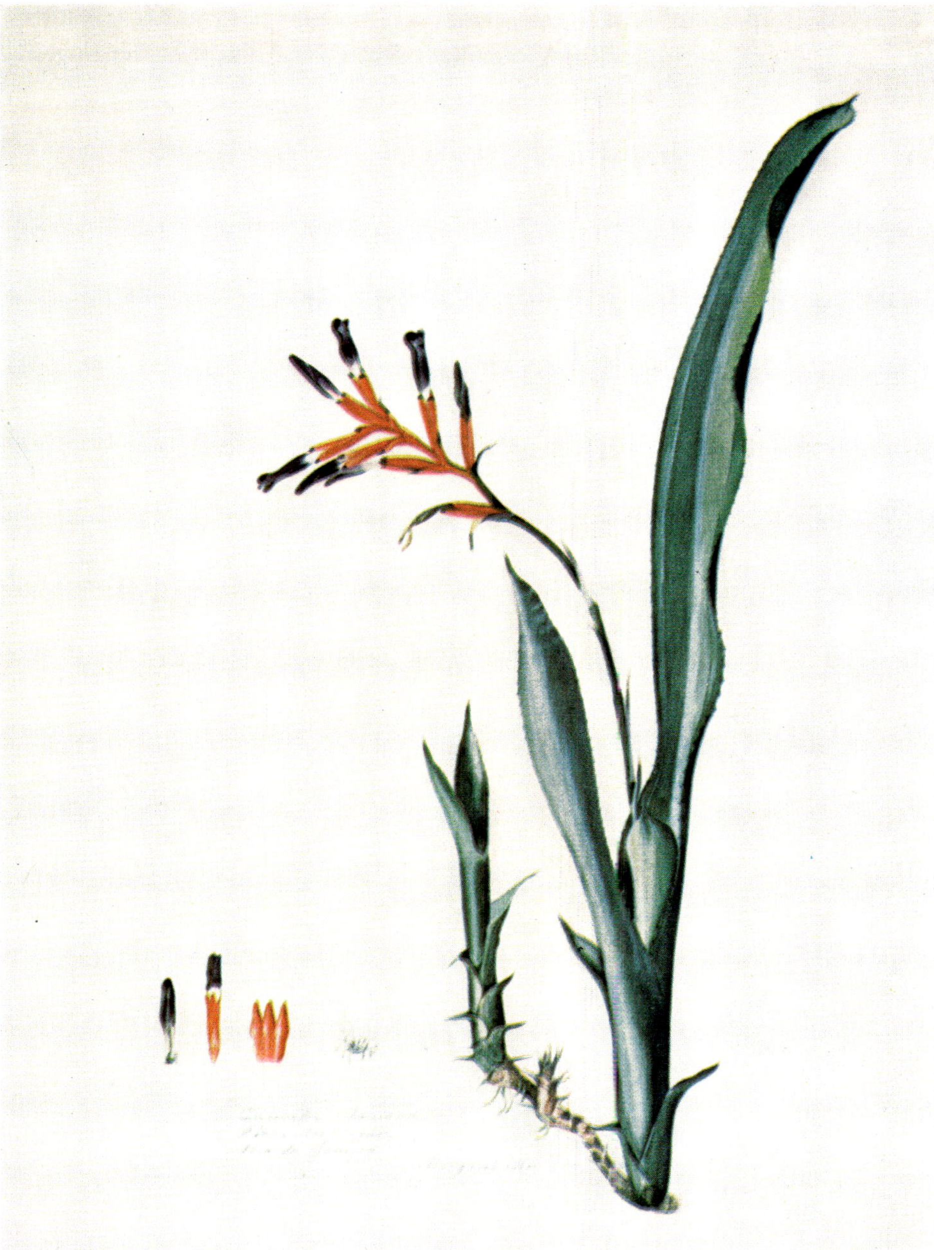

29. *Quesnelia liboniana*

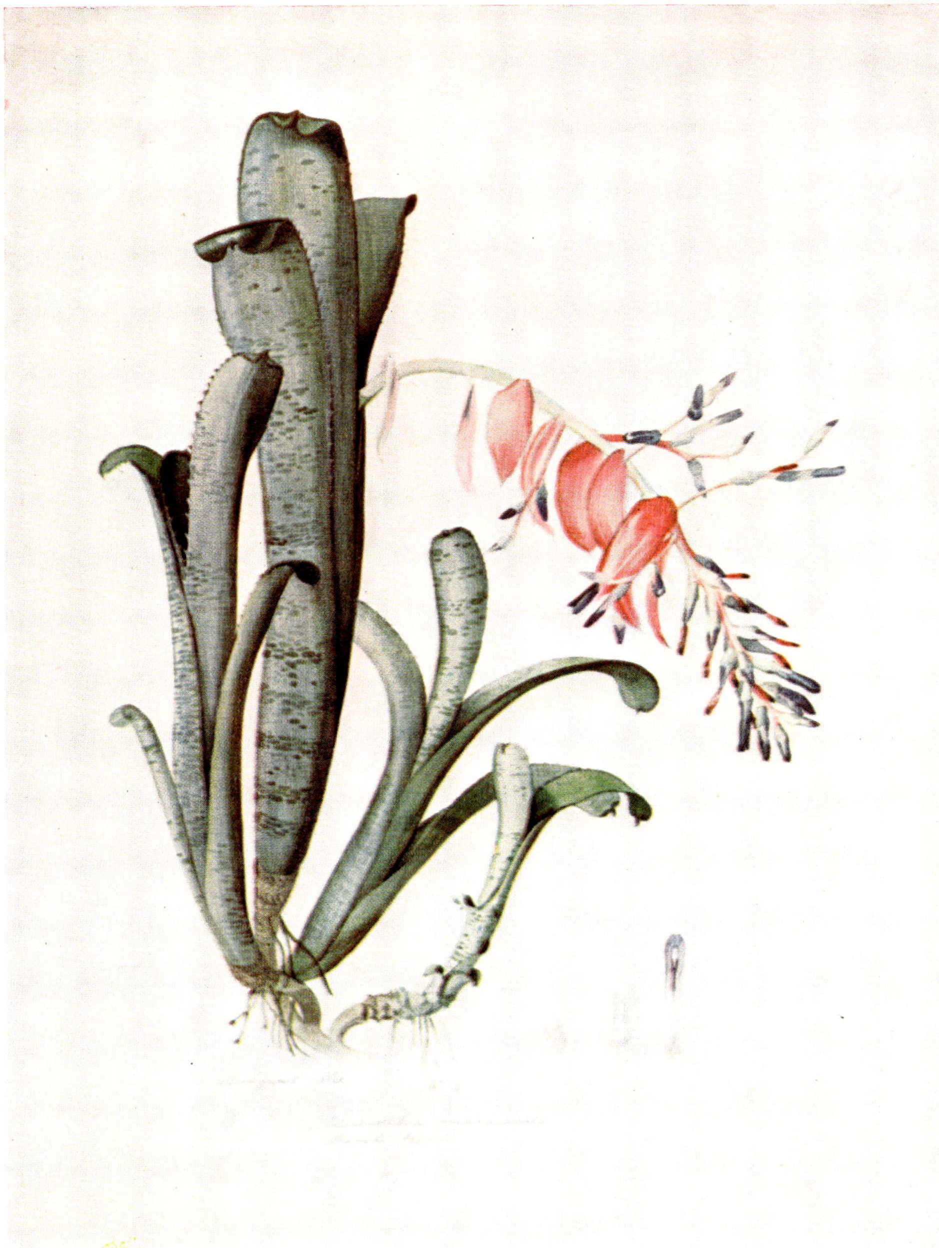

30. Quesnelia marmorata

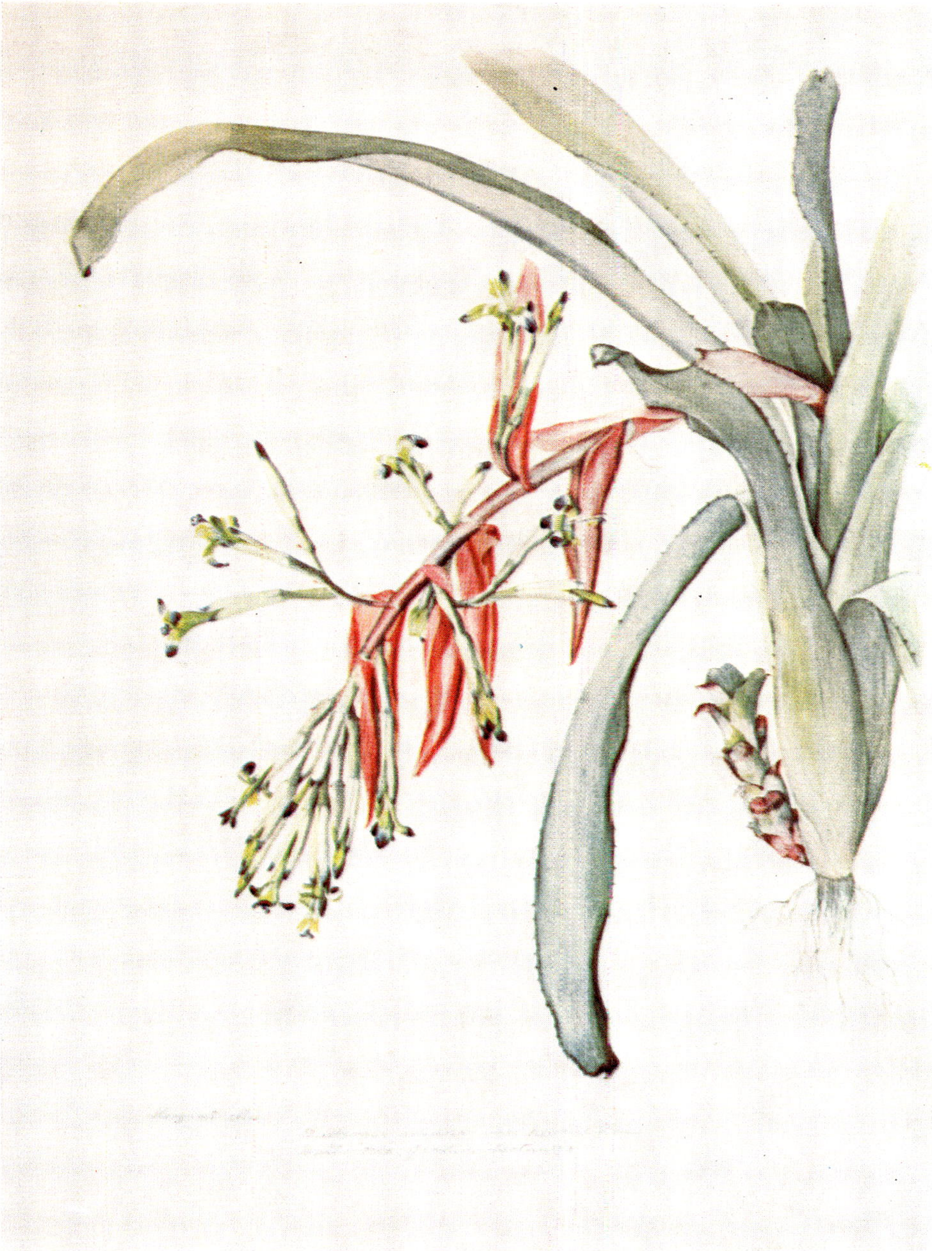

31. *Billbergia amoena* var. *penduliflora*

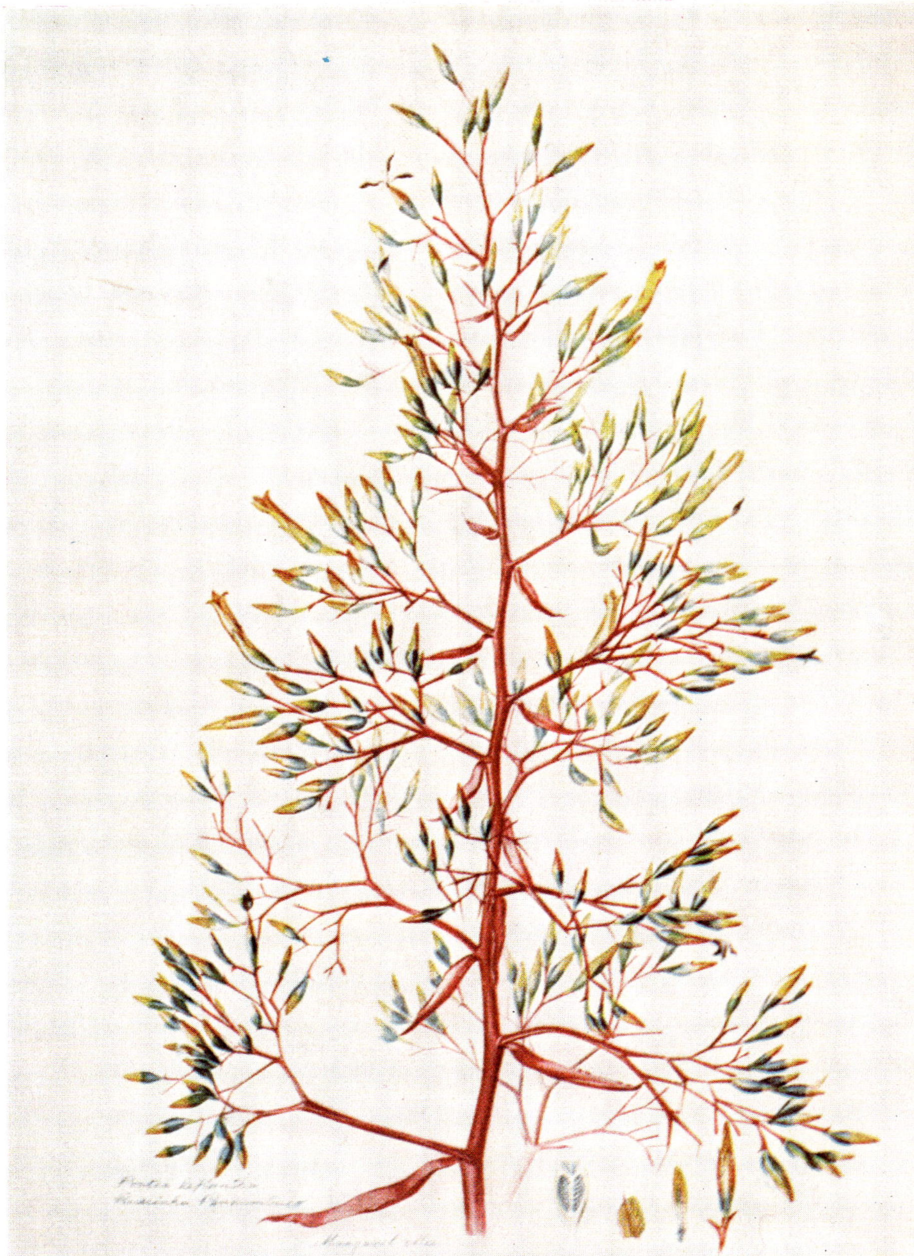

32. *Portea leptantha*